GPI Membrane Anchors – The Much Needed link

eBooks End User License Agreement

Title:

GPI Membrane Anchors – the much needed link

Subtitle:

A biological, structural, physical and chemical look at glycosylphosphatidylinositol-anchored proteins and their influences and potential applications in biotechnology and biomedicine

Editors:

John A. Dangerfield and Christoph Metzner

Authors (in order of appearance by chapter):

Vera Michel, David W.L. Ma, Marica Bakovic, Barbara Viljetić, Marija Heffer-Lauc, Gordan Lauc, Frances J. Sharom, Miren J. Omaetxebarria, Felix Elortza, Martin J. Lear, Bastien Reux, Karthik Sekar, Christoph Metzner, Daniel F. Legler and John A. Dangerfield

This E-book is dedicated to all the wonderful wild-life and natural systems on our planet that are being soullessly destroyed by mankind's greed for economic expansion and wealth.

John A. Dangerfield

Cover art by: Martin J. Lear & Hilda S. King

CONTENTS

Foreword   i

Preface   ii

Contributors   iii

CHAPTERS

1. **GPI-Anchored Proteins and Their Cellular Surroundings: Signalling Function
    and Medical Implications of Membrane Microdomains**

    Vera Michel, David W.L. Ma and Marica Bakovic   **1**

2. **Trans-Cellular Mobility of GPI-Anchored Proteins**

    Barbara Viljetic, Marija Heffer-Lauc and Gordan Lauc   **19**

3. **GPI-Anchored Proteins: Biophysical Behaviour and Cleavage by PI-Specific
    Phospholipases**

    Frances J. Sharom   **34**

4. **Proteomic Approaches for GPI-Anchored Protein Analysis**

    Miren J. Omaetxebarria and Felix Elortza   **53**

5. **Chemical Synthesis, Modification and Mimicry of GPI Anchor**

    Martin J. Lear, Bastien Reux and Karthik Sekar   **64**

6. **Surface Engineering of Biomembranes with GPI-Anchored Proteins and its
    Applications**

    Christoph Metzner, Daniel F. Legler and John A. Dangerfield   **83**

    Glossary   **98**

    Subject Index   **103**

FOREWORD

In the early 90's, gene therapists were beginning to get their heads around the concept of targeted delivery as a means to improve safety and efficacy of gene therapy. Retroviral vectors were, at that time, the most promising vehicles for gene delivery, having been successfully used in the first clinical trials. Naturally, the targeting of retroviral vectors was at the forefront of research, in particular at the level of infection as pointed out in one of the first reviews on this area written by Walter Gunzburg and myself and appearing in Human Gene Therapy. Although an important new concept at the time, it all turned out to be a lot more complicated from today's standpoint.

The modification of virus envelopes fascinated John Dangerfield – he saw viruses as boats that could be equipped with off board ATP driven motors or magnetised by the attachment of iron particles. All very science fiction, I thought, a sort of minaturised lego or mechano set enabling functionalised viruses to be steered so that they deliver their cargo of therapeutic genes to distant ports of damaged or defective cells in the body. Of course, the overriding problem was how to attach these molecules to the virus membrane...

Fast forward to today, and we have a book edited by John and former student and colleague Christoph Metzner, about GPI membrane anchors, a molecular tag that helps insert proteins into the plasma membrane of cells and not just anywhere but in so called lipid rafts, agglomerations or islands of lipid that float in the plasma membrane. The book covers a wide variety of aspects of GPI anchors, what they are, their structure, how they function, what roles they carry out, how they can be used as a tool in biotechnology and how similar GPI-like molecules can be created by molecular mimicry. John and Christoph have selected some of the world's foremost authorities on GPI anchors to contribute chapters to this book. These highly skilled authors provide not only reviews of the knowledge that has been gained about GPI anchors over the last decade or two but also protocols on how to isolate and modify such molecules. Thus the book can be viewed as a nuts and bolts engineering manual and cook book – a kind of "all you ever wanted to know" guide. All in all, the authors have made excellent selections showcasing various facets of current GPI research and shown remarkable editorial skills as well as contributing significant progress themselves to this field.

So why start this foreword with ruminations about the targeting of enveloped viruses like retroviruses? Well GPI anchors offer a solution for the insertion of all kinds of proteins into the plasma membrane not only of cells but also of enveloped viruses and that offers a means to attach ATP motors, magnetic particles and almost any other nanomolecule. This technology, which has been termed "virus painting", but can also be viewed as a type of decorating or coating of enveloped virions, holds much promise for the targeting of safer virus vectors and vaccines as well as having uses in the diagnosis of well-known and emerging virus diseases.

Dr Brian Salmons, President and CEO of SGAUSTRIA

PREFACE

Developing technologies over the best part of the last 40 years have allowed scientists to continually re-address the way cholesterol-rich cell membrane microdomains, or lipid rafts, and their protein components are analysed. More recently this accumulated knowledge has started to take shape into substantial structures of understanding. The focus of this work is on glycosylphosphatidylinositol (GPI)-anchored proteins (GPI-AP) as a major component of lipid rafts. As a special feature, detailed lab-protocols are helpfully integrated into the text in order to allow the replication of state-of-the-art experiments.

The first 2 chapters are of an introductory nature. Firstly, Vera Michel and co-authors review the biological relevance of lipid rafts, what their absence can mean in terms of pathologies and introduce them as "the home of GPI-AP" whereas Barbara Viljetić and co-authors then go into more detail on the basics of GPI-AP. This second chapter has an interesting focus on the mechanisms and reasons for different types of protein release, both with and without their GPI-anchor, as well as their uptake and effect on surrounding cells. Taken together, they give a sound foundation of understanding before going onto the subsequent chapters.

In chapter 3, Frances Sharom gives her interesting insight on the biophysics of the way membranes, GPI-proteins, their anchors and cutting enzymes influence each other, with important implications for function of GPI-AP and their releasing enzymes. In chapter 4, Miren Omaetxebarria and co-authors have a strong technical focus as they take a proteomics approach to analysing GPI-AP, which is rounded up with a look at the bioinformatics aspect. This includes information of the analysis of proteins, anchors (glycans) and the definition of GPI-anchor attachment or omega-sites.

Taking another turn in chapter 5, Martin Lear and co-authors introduce the chemistry of GPIs. Natural GPI-anchors are diverse in structure and yields from cells are low, so the chemical synthesis of GPI-mimics is interesting and may be crucial for up-scaling for multiple applications. This synthetic approach discusses far reaching aspects of the topic and is presented in a way that is accessible for the biological audience. Finally, in chapter 6, the editors, together with Daniel Legler, show that cells, viruses and other biomembrane-encompassed particles can be decorated on their surface with GPI-AP, a process known as "GPI painting" which does not require any genetic manipulation of the modified entities. They further discuss the already emerging applications for this technology.

Aside from addressing the potential that GPI-AP have in perspective of today's developing biomedical and biotechnological scene, one of the aims of this book is to draw together the various scientific disciplines within which GPI-AP and related topics are being researched. Each discipline can of course be justified in its own right; however, by enabling a common level of understanding, novelties may be revealed that allow extraordinary progress to be made.

John A. Dangerfield and Christoph Metzner

CONTRIBUTORS

Bakovic, Marica
mbakovic@uoguelph.ca

Department of Human Health and Nutritional Sciences, Animal Science and Nutrition Building, Room 346, University of Guelph, Guelph, ON, N1G 2W1, Canada.
Tel: +1 519 824 4120 53764

Dangerfield, John A.
john.dangerfield
@vetmeduni.ac.at

and

Department of Pathobiology, Institute of Virology, Christian Doppler Laboratory for Gene Therapeutic Vector Developtment, University of Veterinary Medicine, A-1210 Vienna, Austria.
Tel: +43 1 25077 2301

SGaustria (Austrianova Singapore Pte Ltd), #05-518 Centros, 20 Biopolis Way, Singapore 138668, Singapore.
Tel: +65 6779 2932

Elortza, Felix
felortza@cicbiogune.es

Proteomics Core Facility, CIC bioGUNE, Building 801, Technology Park of Bizkaia, 48160, Derio, Spain.
Tel: +34 944 061 315

Heffer-Lauc, Marija
mheffer@mefos.hr

Department of Medical Biology, J.J. Strossmayer University, School of Medicine, J. Huttlera 4, Osijek, HR-31000, Croatia.
Tel: +385 31 512 845

Lauc, Gordan
glauc@pharma.hr

and

Department of Medical Chemistry and Biochemistry, J.J. Strossmayer University School of Medicine, J. Huttlera 4, Osijek, HR-31000, Croatia.
Tel: +385 31 512 870

Department of Chemistry and Molecular Biology, Faculty of Pharmacy and Biochemistry, University of Zagreb, A. Kovačića 1, Zagreb, HR-10000, Croatia.
Tel: +385 1 639 4467

Lear, Martin J.
martin.lear@nus.edu.sg

Department of Chemistry, Faculty of Science and Medicinal Chemistry Programme, Life Sciences Institute, 3 Science Drive 3, National University of Singapore, Singapore 117543.
Tel: +65 6516 3998

Legler, Daniel F.
daniel.legler@bitg.ch

Biotechnology Institute Thurgau (BITg), University of Konstanz, Unterseestrasse 47, CH-8280 Kreuzlingen, Switzerland.
Tel: +41 71 678 50 30

Ma, David W.L.
davidma@uoguelph.ca

Department of Human Health and Nutritional Sciences, Animal Science and Nutrition Building, Room 243, University of Guelph, Guelph, ON, N1G 2W1, Canada.
Tel: +1 519 824 4120 52272

Metzner, Christoph
christoph.metzner
@vetmeduni.ac.at

Department of Pathobiology, Institute of Virology, University of Veterinary Medicine, A-1210 Vienna, Austria.
Tel: +43 1 25077 2330

Michel, Vera
vmichel@uoguelph.ca

Department of Human Health and Nutritional Sciences, Animal Science and Nutrition Building, Room 309, University of Guelph, Guelph, ON, N1G 2W1, Canada.
Tel: +1 519 824 4120 56717

Omaetxebarria, Miren J.
mirenjosu.omaetxebarria
@ehu.es

Department of Biochemistry and Molecular Biology, Faculty of Science and Technology, Bizkaia Campus, University of The Basque Country, Sarriena z/g, 48940, The Basque Country, Spain.
Tel: +34 9460 13274

Reux, Bastien
bastien.reux@hotmail.fr

Department of Chemistry, Faculty of Science and Medicinal Chemistry Programme, Life Sciences Institute, 3 Science Drive 3, National University of Singapore, Singapore 117543
Tel: +65 6516 3998

Sekar, Karthik
pskkarthikorg@gmail.com

Department of Chemistry, Faculty of Science and Medicinal Chemistry Programme, Life Sciences Institute, 3 Science Drive 3, National University of Singapore, Singapore 117543
Tel: +65 6516 3998

Sharom, Frances J.
fsharom@uoguelph.ca

Department of Molecular and Cellular Biology, University of Geulph, Guelph, ON, N1G 2W1, Canada.
Tel: +1 519 824 4120

Viljetic, Barbara
bviljetic@mefos.hr

Department of Medical Chemistry and Biochemistry, J.J. Strossmayer University School of Medicine, J. Huttlera 4, Osijek, HR-31000, Croatia.
Tel: +385 31 512 826

CHAPTER 1

GPI-Anchored Proteins and Their Cellular Surroundings: Signalling, Function and Medical Implications of Membrane Microdomains

Vera Michel, David W.L. Ma and Marica Bakovic

Abstract: Sphingolipid- and cholesterol-rich membrane microdomains, often referred to as lipid rafts, harbour various plasma membrane proteins and can putatively provide a specific signalling platform for these proteins. Rafts were originally discovered as detergent-resistant lipid domains, but widely differing methods of isolation and analysis have led to numerous classifications, and there is still no coherence which further complicates the study of these domains. Subclasses of lipid rafts have been categorized depending on the specific lipid- and protein-composition, such as caveolae, raft-domains rich in caveolin proteins with tissue-specific signalling properties. Some lipid raft proteins are post-translationally modified by the covalent attachment of a glycosylphosphatidylinositol (GPI) anchor. The complexity of the GPI anchor implies a function beyond directing proteins to the plasma membrane, such as cell-cell communication, regulation of protein structure and cleavage, signal transduction and protein targeting. The mechanisms by which GPI-anchored outer membrane proteins may initiate intracellular signalling events are largely unknown; however several protein kinases have been implicated.

The importance of lipid microdomains becomes apparent in the raft localization of numerous proteins involved in the pathogenesis of diseases. These include neurological diseases such as Alzheimer's and prion diseases, immunological diseases such as lupus erythematosus, and cardiovascular disease. Interestingly, the lipid composition of lipid rafts can be altered nutritionally, especially through dietary long chain fatty acids, which could potentially make lipid rafts an attractive pharmacological target.

Here we will analyze the signalling properties of lipid raft proteins and discuss their importance in health and disease.

MEMBRANE RAFT MICRODOMAINS

The original model of the plasma membrane as a fluid mosaic model was described initially by Singer and Nicholson in 1972 [1]. This model envisioned the membrane as a two-dimensional lipid bilayer with a random distribution of mainly monomeric membrane proteins. While the general principle of a lipid bilayer still holds, it has become obvious since the initial publication of this study that the plasma membrane is a much more complex organelle with a tissue-specific lipid and protein composition, and a distribution of lipids and proteins that is far from being random [2]. Furthermore, the inner and outer leaflet of the bilayer are not the same, but instead provide distinct signalling platforms with differing lipid profiles and specific protein targeting to either layer.

The asymmetry of the membrane bilayer was described in separate studies on erythrocytes in the 1970s and 1980s, where it became apparent that the phospholipid compositions of the inner and outer sheets differ [3-6], and that sphingolipids and cholesterol could possibly play a role in distinct localization of proteins and lipids within the membrane sheets [7,8]. The idea of an asymmetric membrane bilayer culminated in the groundbreaking proposition of the membrane raft model by Brown and Rose, who presented a method to isolate cholesterol- and sphingolipid-rich membrane domains by detergent extraction [9]; the idea of specialized lipid domains floating in the plasma membrane defined the term "lipid raft".

This initial focus on distinct lipid domains was followed by accumulating evidence that the localization of membrane proteins could also be non-random and that proteins are directed to specific areas of the plasma membrane [10,11]. The fact that the formation of rafts could no longer be viewed as an isolated lipid-driven event, but rather a complex interaction of lipids and proteins, led to a recent modification of the term "lipid raft" to "membrane raft" at the Keystone symposium on rafts and cell function in 2006 [12], defining membrane rafts as follows: "Membrane rafts are small (10–200 nm), heterogeneous, highly dynamic, sterol- and sphingolipid-enriched domains that compartmentalize cellular processes. Small rafts can sometimes be stabilized to form larger platforms through protein-protein and protein-lipid interactions." [12]

Despite this agreement on terminology, much disagreement persists as to whether membrane rafts do in fact exist [13,14] and what constitutes a raft, a problem brought about by the high variation in experimental approaches to analyze lipid and protein association with rafts (Table **1A** and **B**). Rafts were first isolated as detergent-resistant membrane domains (DRM) rich in sphingolipids and GPI-anchored proteins floating like rafts on cell preparations subjected to detergent extraction [9]. Although this method is commonly used to study lipid rafts, it has received criticism since treatment of model membranes with detergents such as Triton X-100 has been shown to actually induce the formation of raft-like domains and therefore raises the question if lipid rafts really exist *in vivo* [13,15]. It is now widely accepted that distinctive membrane domains rich in cholesterol and sphingolipids exist; however, depending on the methods employed they are not necessarily considered identical with DRM [16] (Table **1A**). Additional terms, such as the liquid-ordered phase and caveolae, have caused further confusion as to characterize lipid rafts (caveolae will be discussed in Section 2).

Table 1A: Subtypes of Membrane Rafts.

Term	Specific characteristics	Membrane type	Specific lipids and proteins	Reference
Lipid rafts or membrane raft	General term for sphingolipid- and cholesterol-rich cell membrane domains	*In vivo*/live cells	Cholesterol, sphingolipids	12, 26
Detergent-resistant membranes (DRM)	Are only isolated by detergent extraction at 4°C	Live cells or model membranes	Cholesterol, sphingolipids	9, 16
Caveolae	Membrane invaginations rich in caveolins	*In vivo*/live cells	caveolins	52, 54
Liquid-ordered phase	Domains with raft- like characteristics isolated separately from the liquid-disordered and solid phase	Model membranes	Distinct movement in the membrane	17, 18, 19

In vitro membrane models with a known lipid composition, both monolayer [17] and bilayer [18,19], were developed to analyze cholesterol-, sphingolipid- and phospholipid-interactions in these membranes. The membranes typically contain two (binary systems [19]) or three (ternary systems [20]) different membrane lipids that are matched with their fatty acid chain length in order to form a membrane structure. Computer simulation of membrane formation based on different lipid characteristics is also a typical theoretical membrane model system [21]. A systematic classification to characterize different membrane fractions biophysically includes the liquid-ordered phase (l_o) to which lipid rafts are attributed, the liquid disordered phase (l_d) and the solid ordered phase (s_o). Altering the fatty acyl chains of phospho- and sphingolipids [22] and increasing or decreasing the cholesterol concentration[19] can help gaining insight into the stability of membranes with different lipid composition. The idea that an altered fatty acid and cholesterol composition of the membrane changes the conformation of membrane domains has become increasingly important in light of nutritional guidelines for fat intake, and has stimulated research on a targeted modification of the plasma membrane by lipids derived from food [23-25] (we will explore this topic further below). These methods have elucidated the interactions of lipids in the membrane and have further supported the likelihood of sphingolipid- and cholesterol-rich membrane domains; they have led to the proposition of two general models for the formation of lipid rafts. The first model envisisons an interaction between the amide and carboxyl residues in the sphingolipid headgroups [26], while the space between the bulky sphingolipids is filled by cholesterol, an arrangement stabilized through hydrogen bonds and van der Waals-interactions between the cholesterol 3-OH group and the sphingolipid amide groups [27]. The second model attributes raft formation to the stacking of saturated acyl chains also favouring cholesterol packing [28]. While this review focuses on raft lipids and proteins in the plasma membrane, raft-like structures seem to exist in intracellular membranes as well, such as the Golgi membrane [29], but not the mitochondrial membrane [30].

Table 1B: Methods to Analyze Membrane Rafts.

	Method	Advantages	Problems	Reference
Membrane models	Monolayer model membranes	Interactions of membrane lipids can be studied	May not represent the complexity of plasma membrane; may not be reproducible because real plasma membranes are bilayers	17
	Bilayer model membranes	Interactions of membrane lipids can be studied	May not represent the complexity of plasma membrane	18, 19
Isolation of lipid rafts/ DRM	Detergent extraction and discontinuous sucrose gradient centrifugation at 4°C	Classical isolation of detergent-resistant membrane domains	Do DRM only form because of the centrifugation? Does not work at physiological temperature	9
Methods to analyze raft lipids	Treatment of cells with reagents that destroy sphingomyelin or cholesterol	Good method to study importance of sphingomyelin in cells; cyclodextrin is a very established reagent to study cholesterol in membranes	Sphingomyelin and cholesterol may not only be localized in lipid rafts, therefore extraction may not reflect lipid raft function	32, 33, 34
	Treatment with sphingomyelin or cholesterol synthesis inhibitors	Good method to study the importance of these lipids in metabolism	Impaired sphingomyelin or cholesterol synthesis may not only affect raft function	31, 34
	Chromatography lipids in DRM fractions	Common method to analyze raft lipid composition	DRM may not represent real lipid rafts	22, 131
Methods to analyze raft proteins	Co-localization with raft marker in confocal microscopy	Relatively easy and inexpensive	Antibodies may interfere, or marker antibody may induce artificial clustering of rafts	33, 35, 67, 68, 69
	Co-localization with marker in isolated DRM	Relatively easy and inexpensive	Questionable if raft-like domains form because of this isolation method	66, 70, 72
	Transfection of cells with caveolin (other proteins) small interference RNA	Very good method to study impaired protein expression in tissue culture cells	Relatively complex procedure, development of siRNA, may not completely inhibit protein expression	63, 64
	Knockout or transgenic mouse models to study the loss or over-expression of raft marker proteins	Very good method to study importance of raft resident proteins *in vivo*	Very complex, expensive and long procedure	65

The theoretical membrane models can not accurately mimic the complexity of the plasma membrane; however, numerous studies in living cells have further established that cholesterol and sphingolipids are the most important components of raft domains. A common approach is to disrupt the synthesis of these compounds or extract them from the membrane and then observe altered metabolism of the cell. For example, HMG-CoA-reductase, an enzyme involved in cholesterol synthesis, can be inhibited by mevinolin [31], and cholesterol extraction can be induced by treatment with β-cyclodextrin [32,33], which are both common approaches to study the importance of lipid rafts in membrane protein signaling. Fumonisin B inhibits sphingolipid synthesis, while the enzyme sphingomyelinase disrupts membrane sphingomyelin [34]. Alternatively, cells can be treated with a fluorescent compound that specifically interacts with raft lipids and can then be visualized [35]. For example, the pore-forming protein lysenin interacts with sphingomyelin, while polyethylene glycol cholesterol esters bind cholesterol-rich membrane domains [35]. The biggest downfall of these studies is the complexity of the problem: disruption of cholesterol metabolism, for example, has an impact on the overall membrane structure and makes it challenging to separate changes in raft structure caused by such a disruption from an overall depletion of cell cholesterol. Table **1B** summarizes common methods for lipid raft analysis. Recent advances in methodology, such as fluorescence energy transfer (FRET) microscopy, immunoelectron microscopy and single-particle tracking, may aid in clarification of the nature of lipid rafts in live cells.

Despite the lack of agreement on membrane raft isolation and characterization, there is no doubt that the plasma membrane is a complex organelle with distinct lipid and protein signalling platforms, and that there is an ordered arrangement of proteins in specific lipid domains which differs between the inner and outer membrane sheets.

PROTEIN LOCALIZATION IN LIPID RAFTS: SIGNALLING AND FUNCTION

Protein Localization to Rafts

Numerous proteins with abundant functions such as nutrient and ion transport, membrane sorting, membrane trafficking, signal transduction and cell polarization, have been associated with membrane rafts. They can be roughly categorized into two groups: raft resident proteins which permanently localize to rafts, such as the caveolins, and proteins which only localize to rafts upon activation and initiation of their function, such as the T-cell receptor. Distinct mechanisms seem to induce partitioning of proteins to membrane rafts. These include the glycosylphosphatidylinositol (GPI)-anchor, which targets proteins to rafts of the outer membrane leaflet, and a single or multiple fatty acylation, which seems to localize proteins to raft domains of the inner membrane.

The GPI-anchor is a complex glycolipid structure which is post-translationally added to the C-terminus of many cell-surface proteins. Proteins destined for this modification contain both a C-terminal sequence signalling addition of the GPI-anchor in the endoplasmic reticulum (ER), as well as an N-terminal ER-targeting signal. The C-terminal GPI signal sequence (GSS) is cleaved off in the ER and replaced by the GPI-anchor, a reaction catalyzed by the enzyme GPI-transamidase. The anchor is composed of a phosphoethanolamine linker connecting the anchor to the protein, a glycan core with varying sugar side chain additions, and a phospholipid tail which ensures membrane insertion [36]. GPI synthesis takes place in the ER in a multi-step process involving the PIG (phosphatidylinositol glycan anchor synthesis) enzymes. Initially, N-acetyl-glucosamine (GlcNac) is transferred to the phosphoethanolamine moiety of phosphatidylinositol (PI) and then deacetylated to glucosamine (GlcN) at the cytoplasmic side of the ER. PI features a saturated fatty acid side chain at the sn-1 position and an unsaturated fatty acid side chain at the sn-2 position. This initial transfer is catalyzed by a six-enzyme complex called GPI-GlcNac transferase, with the enzyme PIG-A representing the catalytic unit. The PI-GlcN complex then flips to the luminal side of the ER, where PIG-W (an acyltransferase) mediates the addition of an acyl chain to the PI-inositol ring from acyl-CoA to form acyl-PI-GlcN. In subsequent steps, three mannose residues are transferred from Dol-P-Man to the GlcN by the enzyme complex PIG-M/PIG-X [37], and phosphoethaolamine residues are attached to each mannose by the enzymes PIG-N, PIG-O and PIG-F. The final phosphoethanolamine residue on mannose 3 connects the GPI-anchor to the protein. Before the GPI-linked protein is transported to the Golgi network through the secretory pathway, the acyl-chain is removed from the PI-inositol ring by the enzyme post-GPI attachment to proteins (1PGAP1). In the Golgi, the unsaturated fatty acid side chain in position sn-2 of PI is replaced by a saturated fatty acid through the enzymes PGAP2 and 3. Mature GPI-anchors therefore contain two saturated fatty acid side chains, which likely facilitates their positioning in the plasma membrane. It is in the Golgi that GPI-linked proteins assemble in raft like structures and are thereby targeted to the cell surface [38,39].

Acylation refers to either the co-translational addition of myristate, a C14 saturated fatty acid, to a glycine near the protein's N-terminus, or a posttranslational addition of a longer fatty acid to a cysteine residue of the protein. This S-acyl chain is often a palmitate but could also be a longer and/or unsaturated fatty acid. It is likely that such a modification would not be the only localizing factor, but that protein-protein and protein-lipid interactions within membrane rafts also ensure protein localization, and acylated proteins are often both myristoylated and palmitoylated. Some proteins are known to interact with raft lipids, and this interaction seems to involve the transmembrane domains of a protein and/or the N-terminus [40]. A distinct example of protein-lipid interaction is that of caveolin, which binds directly to raft cholesterol and thereby ensures its resident raft localization [41].

An attractive aspect of plasma membrane compartmentalization into rafts is that it presents a specific platform for localization of receptor proteins and a regulated activation of signalling cascades. Countless signaling proteins have been associated with rafts in one way or the other, including receptor proteins such as the estrogen receptor [42] and growth factor receptors [43], ion channels such as Ca2+ pumps [44,45] and ATPases [46,47], kinases such as protein kinases A [48] and C, PI3K and MAPKinases, as well as other downstream effectors such as G proteins and eNOS. We will exemplify diverse mechanisms for raft protein signalling in health and disease in section III.

Raft Resident Proteins

Caveolae are a subclass of lipid rafts which are flask-shaped invaginations of the cell membrane and contain caveolin proteins. Three isoforms of caveolins can be distinguished [49], caveolin-1, -2 and -3, which are transcribed from different genes. The genes for caveolin-1 and -2 are localized on chromosome 7, while the caveolin-3 gene is on chromosome 3. Caveolae rafts are thought to form from smaller lipid raft entities upon multimerization of caveolae molecules; an oligomer of 15 caveolin molecules seems to be the minimum required for caveolae formation [50], however caveolae may contain several hundred caveolin molecules and are considered a rather large raft entity with an average diameter of ~100 nm. Caveolae have been implicated in signal transduction, cholesterol transport and exocytosis, and importantly present a distinct platform for a specialized type of endocytosis.

The classic mechanism for the uptake of extracellular molecules and plasma membrane proteins is clathrin-mediated endocytosis, where assembly of clathrin molecules induces the formation of coated vesicles which are subsequently invaginated inside the cell and then transported to the endosome for degradation or recycling [51] (Fig. **1**).

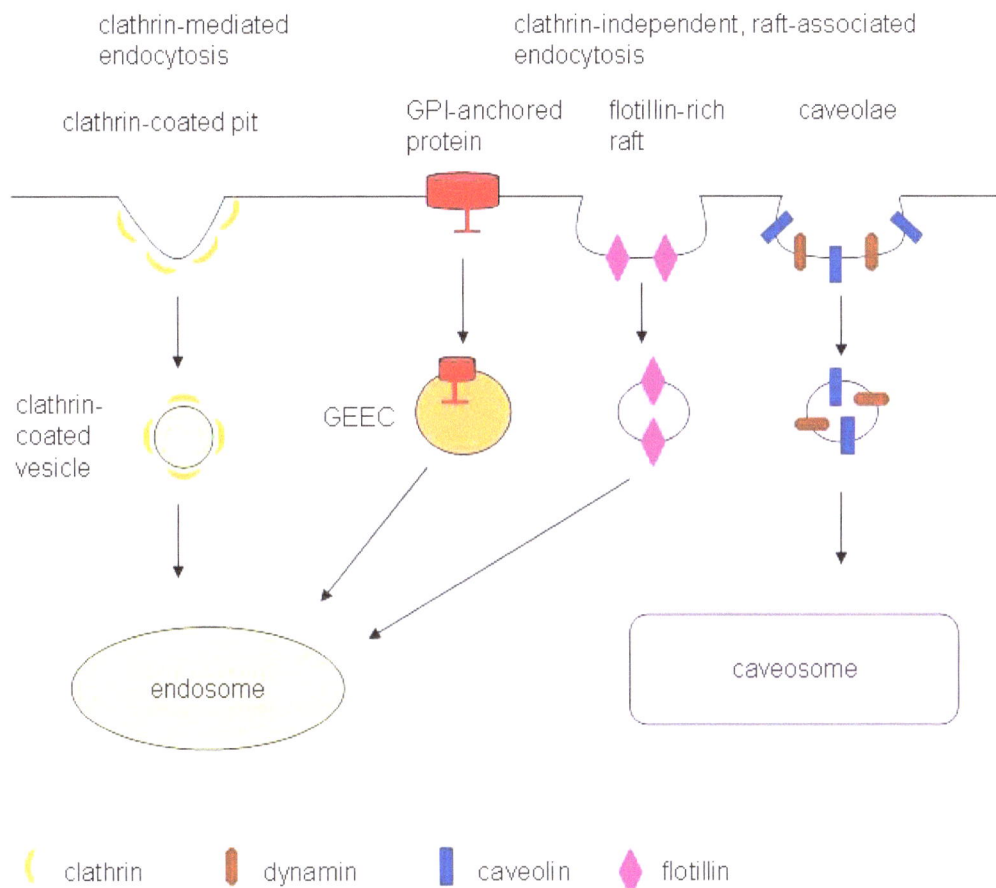

Figure 1: Endocytosis mechanisms. Clathrin-mediated endocytosis: Clathrin polymerizes at the plasma membrane forming a coated pit, which is then internalized as a clathrin-coated vesicle and transported to the endosome, from where it is either recycled to the plasma membrane or passed on to the lysosome for degradation. GPI-anchored proteins: GPI-proteins are endocytosed in GEEC vesicles, a process regulated by small GTPases and the protein GRAF1. Flotillin-associated endocytosis. Flotillin-rich vesicles are internalized from the plasma membrane and transported to the endosome in a clathrin-independent mechanism. Caveolae-mediated endocytosis:In this clathrin-independent pathway caveolae are internalized supported by the protein dynamin and the traffic to intracellular compartments such as the endosome or a caveolae-specific endosome-like organelle, the caveosome.

However, recent studies have shown that there are clathrin-independent mechanisms of endocytosis involving different proteins which are all associated with lipid rafts [52] and can be inhibited by cholesterol depletion [53,54]. Some of these pathways are dependent on caveolae and/or caveolin-1 [55], others seem to be related to raft domains lacking caveolin proteins [56]. The discovery that caveolae can bud from the membrane and traffic in vesicles inside the cell [57] and that dynamin, a protein that can induce formation and detachment of membrane vesicles, is located in lipid rafts55 further supports a role of these membrane domains in endocytosis. Several pathogens use caveolae to mediate their uptake into the host cell, particularly epithelial cells, such as the viruses simian virus SV-40 [58] and the influenza virus [59] as well as the bacteria *Campylobacter jejuni* [60] and *Escherichia coli* [61]. Although caveolin proteins have been shown to be required for caveolae formation and stabilization [62,63], their specific role in caveolae-dependent endocytosis is unclear. It has been proposed that they may even impair caveolae-vesicle formation as caveolin-1 expression is inversely related to caveolae budding [64]. This implies that either caveolin-1 is actively involved in the regulation of caveolae-dependent endocytosis, possibly by phosphory-lation of the protein itself [53], or that caveolae may not be as tightly interconnected with caveolins as previously thought. However, intracellular caveolin-containing vesicles have been shown to traffic from the cell membrane inside the cell [57]. Therefore, the involvement of caveolins in this type of endocytosis remains to be elucidated. The knowledge of the intracellular trafficking of clathrin-coated vesicles to a sorting compartment, the endosome, raised the question if there is a similar compartment specific for the uptake of caveolae-vesicles. Such a compartment, named the caveosome, has been identified as an organelle distinct from the classical endosome to which caveolin-positive vesicles seem to traffic [57,65].

Aside from caveolae, there may be other clathrin-independent, raft-associated pathways of endocytosis. One recently characterized mechanism involves another raft resident protein, flotillin-1. Glebov *et al.* [56] identified budding of flotillin-rich membrane regions which where different from caveolae domains. This mechanism was clathrin-independent as shown by a lack of co-localization with transferrin, a typical marker of clathrin-mediated endocytosis, in confocal microscopy immunofluorescence studies in HeLa cells. Furthermore, these flotillin-rich vesicles localized intracellularly to an endosome compartment that also contained GPI-linked proteins as well as cholera toxin B subunit, both typical raft resident proteins. The uptake of these proteins from the plasma membrane was inhibited by tranfection of HeLa cells with flotillin-1 small interference RNA. These results demonstrate a clathrin- and caveolae-independent, flotillin-1-dependent new pathway of endocytosis associated with lipid rafts (Fig. **1**).

GPI-anchored proteins seem to undergo an endocytotic pathway distinct from both the caveosome and the clathrin-coated pits. They are endocytosed through clathrin-negative, GPI-anchored protein enriched early endosomal compartments (GEEC); this endocytic pathway is regulated by GTPases through the GTPase regulator associated with focal adhesion kinase 1 (GRAF1) protein [66]. It is as yet unclear why and how the GPI-anchor confers them to special endosomes, however, a recent study proposes that this is due to not the structure of the anchor itself, but instead steric limitations due to its conformation [67].

The establishment of raft marker proteins such as caveolins and flotillins can be used to study membrane rafts by detection and modification of their expression. Caveolin expression can be decreased by transfection of cells with caveolin small interference RNA (siRNA [68,69]) or knockout of the caveolin gene in *in vivo* mice models [70]. Alternatively, caveolin expression can be increased by introduction of caveolin cDNA vectors into the cell and subsequent overexpression of the gene [71]. The localization of other proteins to lipid rafts can be demonstrated by disruption of raft domains and/or inhibition of caveolin expression and subsequent changes in their function. Co-localization of proteins and lipid rafts can be confirmed by confocal microscopy, where the protein of interest is identified with a specific fluorescent-labeled antibody and the raft fraction with an antibody against caveolin [72], flotillin-133, [73] or the cholera toxin B subunit [74].

****

PROTOCOL 1

Gas Chromatographic Analysis of Fatty Acids:

1. *100 µl aliquot of membrane raft is transferred to a 15 ml glass screw cap tube (Pyrex) with Teflon cap.*

2. *Lipids are extracted in chloroform:methanol 4 ml (2:1) and KCl 1 ml (0.88%w/v), vortex.*

3. *Flush tube with nitrogen and sit overnight.*

4. *Centrifuge at 200 x g to separate phases.*

5. *Carefully transfer bottom chloroform fraction to a clean glass tube, with Teflon cap.*

6. *Add 2 ml of chloroform and repeat extraction from step 4.*

7. *Combine chloroform extracts and dry down under a gentle stream of nitrogen.*

8. *Saponify lipids by adding 2 ml of KOH-MeOH (0.5M), then incubate at $100^{\circ}C$ (water bath, sand bath or oven) for 1 hr, then cool for 5 min in fume hood.*

9. *Methylate fatty acids by adding 2 ml of 14% BF_3-MeOH and 2 ml hexane, then incubate at $100^{\circ}C$ (water bath, sand bath or oven) for 1 hr.*

10. *Cool for 5 min in fume hood and then add 2 ml of double distilled water.*

11. *Centrifuge at 200 x g to separate phases.*

12. *Transfer upper hexane phase containing fatty acid methyl esters (FAME) to clean 2 ml gas chromatography vials.*

13. *Dry down hexane under gentle stream of nitrogen.*

14. *FAME are quantified on an Agilent 7890 flame-ionisation detector gas chromatograph (Palo Alto, CA, USA) separated on a SP-2560 (Supelco) fused silica column, 100m x 0.25mmID x 0.20µm. Samples are injected in splitless mode. The injector and detector ports were set at $250°C$. FAME are eluted using a temperature program set initially at $60°C$, increased at $13°C/min$ and held at $170°C$ for 4 min, increased at $6.5°C/min$ to $175°C$, increased at $2.6°C/min$ to $185°C$, increased at $1.3°C/min$ to $190°C$, then increased at $13°C/min$ and held at $240°C$ for 13 min. Total run time is approximately 35 min. The carrier gas, hydrogen, is set to 0.5 ml/min constant flow rate. The makeup gas is nitrogen. Fatty acids are identified using authentic standards obtained from Nu-Chek Prep.*

****

MEDICAL IMPLICATIONS OF RAFT PROTEIN LOCALIZATION

Immunological Disease

One of the probably best described GPI-linked proteins is Thy-1/CD90, a small lipid raft protein of immune and neuronal cells involved in diverse signalling pathways. T-cell activation is initiated by the binding of antigen to the T-cell receptor (TCR) and a subsequent interaction of a co-activator protein (CD4 or CD8) with the MHCII or MHCI complex of the antigen-presenting cell (extracellularly) and the tyrosine kinase Lck (intracellularly). Upon antigen binding, T-cell receptors dimerize and the immunoreceptor tyrosine-based activation motive (ITAM) in the intracellular domain of the TCR is phosphorylated by Lck. Phosphorylation of the TCR recruits the tyrosine kinase ZAP to the membrane, which in turn phosphorylates the adapter molecule "linker for activation of T-cells" (LAT). Phosphorylated LAT initiates the assembly of a multiprotein complex which signals the activation of transcription and actin remodelling. Evidence has accumulated for the localization of T-cell activation in lipid raft domains: the T-cell receptor (TCR) is thought to localize to lipid rafts upon antigen stimulation, and its co-receptor CD4 and the TCR-activating tyrosine kinase Lck are both acylated and also present in rafts. Furthermore, cross-linking of the typical raft lipid GM1 can stimulate T cell activation. Thy-1 has been suggested as a T-cell activator protein in a fashion distinct from the T-cell receptor [75], and Thy-1 downstream signalling was recently linked to intracellular activation of JNK and ERK kinases and subsequent interleukin-2 production [76].

Systemic Lupus erythematosus (SLE) is an autoimmune disease characterized by the development of autoantibodies against nuclear antigens and abnormal T-cell signalling, subsequently leading to inflammatory damage of multiple organs. Hyper-activation of the TCR is believed to be involved in the pathogenesis of SLE, however the precise mechanisms are currently unknown. Recent studies have demonstrated that disruption of lipid rafts slows SLE disease progression, while induced aggregation of lipid rafts seems to have the opposite effect. T-cell activation in these patients seems to be closely linked to an elevated amount of raft lipids in the plasma membrane, with a higher GM1 content and a subsequently increased number of lipid rafts [77,78]. Downstream from the TCR, abnormal expression and/or activation patterns of the above mentioned proteins in the TCR signalling cascade are well documented. For example, Lck expression is reduced in SLE activated T-cells, and Lck raft localization is decreased. The low amount of Lck still present in lipid rafts is primarily in the activated, phosphorylated form [79].

Neurological Disease

The neuronal GPI-linked cell adhesion molecules NCAM and Thy-1 both induce intracellular signalling cascades and provide examples of GPI-mediated cell-cell communication [80]. Lipid raft-localized neural cell adhesion molecule (NCAM) signals through the non-receptor tyrosine kinase fyn by induction of fyn autophosphorylation, which then binds and phosphorylates the focal adhesion kinase fak [81]. Fak in turn activates the G-protein Ras and further downstream the ERK kinase. Neuronal Thy-1 is involved in synapse stability and neurite outgrowth and, similarly to NCAM, signals through Fak to mediate cell adhesion; importantly, the GPI-anchor appears to be critical to mediate these functions [82].

Prion diseases are characterized by a pathological conversion of the cellular glycoprotein PrPc (= Prion-related Protein; encoded by Prnp gene on chromosome 20) to the infectious form PrPSc, which causes neurodegenerative diseases referred to as bovine spongiphorm encephalopathy (BSE) in cattle, scrapie in sheep and Creutzfeld-Jacob disease (CJD) in humans. The disease progression in characterized by PrPSc aggregation and accumulation in the brain and a sponge-like degeneration of brain tissue [83], which causes irreversible brain damage and is therefore always fatal. The physiological function of cellular PrP is not well characterized; the protein is primarily present in neural and immune cells and is thought to be involved in copper metabolism due to a C-terminal Cu^{2+} binding domain. Furthermore, it has been implicated to play a role as an antioxidant [84], in apoptosis [85], cell proliferation and in signal transduction [86]. The infectious PrPSc is characterized by an increase in β–sheets in the C-terminus of the protein which normally forms mostly α–helices in PrPc, which renders PrPSc resistant to protease-mediated cleavage. Both PrPc and PrPSc have been shown to associate with rafts [87] and there is some indication that the conversion takes place within rafts [88]. PrP is a GPI-anchored protein and the GPI-anchor is required for raft localization [31]; in addition, PrPc seems to have an N-terminal signal which also confers localization to rafts [89]. Depletion of cholesterol with mevinolin impairs PrPc localization to the plasma membrane and leads to accumulation of PrP in the Golgi31, while depletion of sphingolipids seems to have no such effect [90]. The subclass of rafts to which PrP localizes seems to be devoid of caveolins, however PrPSc can alter caveolin-1 raft distribution and seems therefore to be indirectly connected to caveolin signalling in neural cells [91]; PrP rafts are furthermore unusually rich in hexosylceramide and unsaturated fatty acids compared to for example Thy-1 rafts [92]. Interestingly, PrP is internalized by a type of endocytosis which is both raft- and clathrin-dependent but caveolin-1 independent [93,94].

Alzheimer's disease (AD) is a neurodegenerative disease manifested by the aggregation of plaques in the brain causing detrimental morphological and functional changes. The plaques consist primarily of the amyloid-β-peptide (A beta), which results from the proteolytic cleavage of the integral membrane protein amyloid precursor protein (APP) mediated by the enzyme beta-secretase. This amyloidogenic processing of APP is thought to occur within lipid rafts based on the findings that beta secretase localizes to these domains, and that A beta aggregation seems to be connected to cholesterol [95] and lipid rafts [96] and gangliosides GM1 and GM2 [97]. Cholesterol depletion inhibits A beta formation in hippocampal neurons [98], while model membranes with an increased cholesterol content show an elevated accumulation of GM1-A beta clusters [99]. This finding goes along with the observation that GM1 and GM2 concentrations are increased in AD brains [100].

Parkinson's disease (PD) is another neurogenerative disease manifested by the destruction of dopamine-producing neurons in the substantia nigra brain region leading to a decrease in dopamine production and signalling [101]. The loss of dopamine signalling causes a loss of control over movement and leads to an uncontrollable muscle contraction [102]. Other symptoms include cognitive disorders such as dementia [103], depression [104] and anxiety [105]. Mutation [106] and aggregation [107] of a protein with as yet unknown function present in both healthy and PD brains, alpha synuclein, has been associated with PD pathogenesis [108]. Several lines of evidence point to the raft localization of alpha-synuclein:

Overexpression of alpha-synuclein in neuroblastoma cells caused an upregulation of caveolin-1 expression, followed by an impaired signalling of the MAP kinase pathway [109]. Furthermore, alpha synuclein has been shown to directly interact with raft regions [110], likely in a lipid-mediated association with rafts [110,111]. Another raft resident, flotillin-1, shows up-regulated expression in do dopaminergic neurons of PD brains [112].

Cardiovascular Disease

Both acquired and inherited GPI-deficiencies seem to induce thrombosis and intravascular complications. PNH (paroxysmal nocturnal haemoglobinuria) is an acquired disease caused by a mutation of the enzyme catalyzing the first step in GPI synthesis, PIG-A [113]. The loss of the GPI-anchor primarily affects the expression of the GPI-linked lipid raft protein CD59, which functions as a complement activator inhibitor in erythrocytes. CD59 signalling has been linked to activation of the phosphotyrosine kinase Hck, which subsequently tyrosine phosphorylates downstream adapter proteins and kinases [114,115]. The disease is characterized by intravascular haemolysis, iron-deficiency and anaemia, thrombosis and haemorrhage. Life expectancy of PNH patients is quite low, and acute treatment involves iron supplements to treat the anaemia and anticoagulants to prevent thrombosis.

There have been a few reports on patients with inherited GPI-deficiency, and this disease is caused by a loss of PIG-M expression, the enzyme which transfers mannose residues to the growing GPI-anchor. The decreased PIG-M expression was recently shown to be caused by a mutation in the PIG-M promoter [116].

Caveolin-3 is exclusively expressed in muscle cells, including the sarcolemma of cardiomyocytes [117]. Caveolin-3 signaling has been linked to cardiac hypertrophy, a compensatory mechanism characterized by the enlargement of the heart to accommodate increased energy demand. Hypertrophy can be physiological during a stress such as exercise, however can be a manifestation of serious conditions such as ischemia and hypertension and lead to abnormal heart function. Induction of hypertrophy by the alpha-adrenergic agonist phenylephrine in neonatal rat cardiomyocytes led to an up-regulation of caveolin-3 expression and an increase in caveolae number [118]. However, treatment of the rat cardiomyoblast cell line H9c2 with phenylephrine had the opposite effect: Caveolin-3 expression and localization in caveolae was decreased [119]. This is in agreement with observations from a caveolin-3 knockout mouse model, which develop cardiac hypertrophy and therefore seem to imply an opposite regulation of caveolin-3 expression and hypertension [70,120]. It is possible that these discrepancies result from a difference in developmental stage of the cell model [121]; evidence points towards an inverse relation between caveolin-3 and hypertrophy in adult cells. How does caveolin-3 impair cardiomyocyte growth? Caveolin-3 has been shown to directly interact with and inhibit various growth factor receptors [122,123], and the inhibition of hypertrophy seems to be achieved by and impairment of growth signals; a recent study demonstrated a Ca2+-dependent caveolin-3 mediated growth inhibition in H9c2 cardiomyoblasts [123]. The downstream mediators of this inhibition are not well characterized; caveolin-3 knockout mice show a hyperactivation of MAPK42/44 [70], and overexpression of caveolin-3 has the opposite effect [123,124]. PKC, small GTPases and p38 MAPK have also been shown to be downstream targets of caveolin-3; the respective cascade seems to be specific to the signal that triggered the hypertrophy.

Several studies aimed to link hypertension to lipid raft signalling have investigated receptor-mediated signaling in endothelial cells of arteries and the heart muscle. The receptor for angiotensin II (Ang II), a peptide that stimulates vasoconstriction and is capable of inducing hypertension, associates with lipid rafts in vascular smooth muscle cells upon angiotensin II binding [69]. The receptor activation initiates in turn the activation of the G-protein Rac I and its translocation to the plasma membrane (Fig. **2**). NAD(P)H oxidase [125], a membrane-bound enzyme that catalyzes the formation of reactive oxygen species (ROS) such as H2O2 [126] is activated by GTP-Rac I; ROS can then trigger signaling cascades such as the translocation of the epidermal growth factor receptor (EGF) to the plasma membrane and ensuing downstream MAP kinase and Akt signalling [69]. Caveolin-1 knockdown prevents the localization of the AngII receptor to rafts and impairs Rac I GDP-GTP exchange and its recruitment to the plasma membrane [69]. The receptor-mediated translocation of activated G-proteins to rafts is also observed in cardiac myocytes for adrenergic [127,128] and cholinergic [129] receptors.

As becomes apparent from the above mentioned interdependency of Ca2+ and caveolin-3 signaling, caveolae are crucially involved in cardiac ion channel function; various ion channels localize to caveolae in the heart, including calcium channels, a sodium channel [130], the Na+-K+-exchanger [131] and numerous potassium channels (Kv1, Kv2, Kv4, Kir2, Kir3, KATP). Since cardiac ion channels regulate membrane potential, vasoconstriction/-dilation, and heart muscle contraction and blood flow, the localization to caveolae underlines the importance of distinct signalling domains provided by caveolin-3 and caveolae and their disruption can cause hypertension, ischemia and heart failure [34].

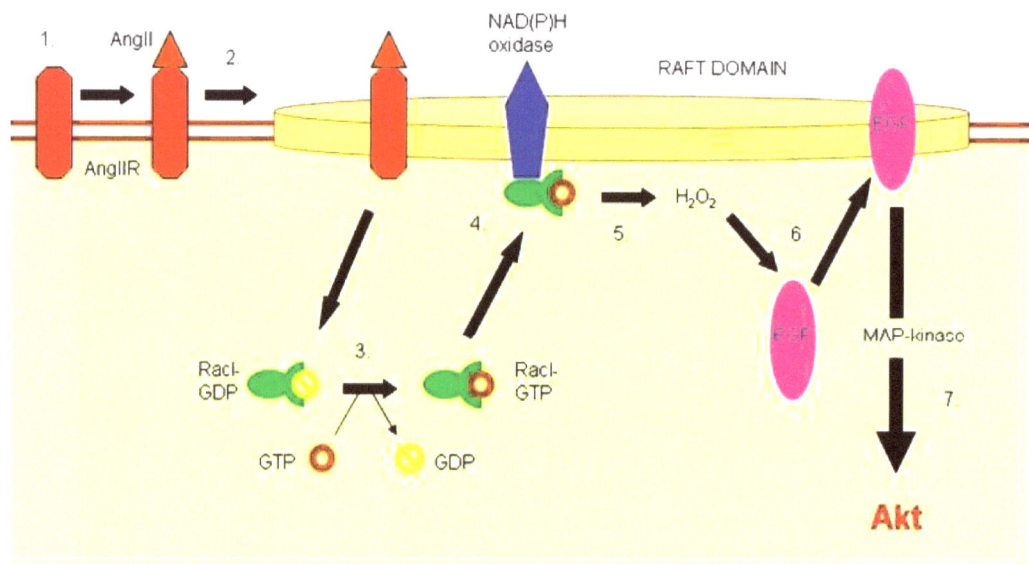

Figure 2: Raft-mediated angiotensin II signaling. Angiotensin II (AngII) binds to the angiotensin II receptor (AngIIR) (1.) which then localizes to lipid rafts (2.). The activated receptor induces the activation of Rac I by exchanging a Rac I-bound GDP against a GTP (3.). Rac I can then activate NAD(P)H oxidase in lipid rafts (4.), which catalyzes the formation of reactive oxygen species such as H2O2 (5.). These compounds can induce the translocation of the EGF receptor to the plasma membrane (6.), which upon EGF binding stimulates downstream cascades involving MAP kinases and Akt (7.).

In contrast to vasoconstriction mediated by angiotensin II, vasodilation and muscle relaxation mediated by nitric oxide (NO) is also lipid raft-associated in cardiac myocytes and endothelial cells. Endothelial nitric oxide synthase (eNOS) generates NO in cardiac myocytes, where it inhibits adenylate cyclase, the enzyme producing cyclic adenosylmonophosphate (cAMP). cAMP is a signaling molecule involved in the regulation of muscle contractility. eNOS and adenylate cyclase both localize to lipid rafts, and disruption of cholesterol synthesis with cyclodextrin leads to impaired AC function32. eNOS also directly interacts with caveolin-1 and this interaction is involved in eNOS regulation [132].

Apoptosis and Carcinogenesis

Apoptosis is a programmed cell death which provides a crucial control of cell growth and regulates the removal of excess cells and thereby cancer proliferation. Depending on the cell type, two different pathways are known to induce this controlled cell suicide: Type I apoptosis is mediated by an extrinsic, receptor dependent pathway and is necessitates the binding of a ligand to a death receptor such as TNF or Fas/CD95. Type II apoptosis is mediated through primarily intracellular, intrinsic pathways involving cell organelles such as mitochondria and/or the endoplasmic reticulum. Caspases are proteolytic enzymes downstream of both pathways which essentially act as the final executors of apoptosis. Fas/CD95 is a death receptor involved in type I apoptosis, and has recently been convincingly localized to lipid rafts in several studies [133,134]. Upon ligand binding, Fas/CD95 trimerizes, thereby recruiting the Fas-associated death domain protein (FADD) and caspase-8 [135,136] (Fig. **3**).

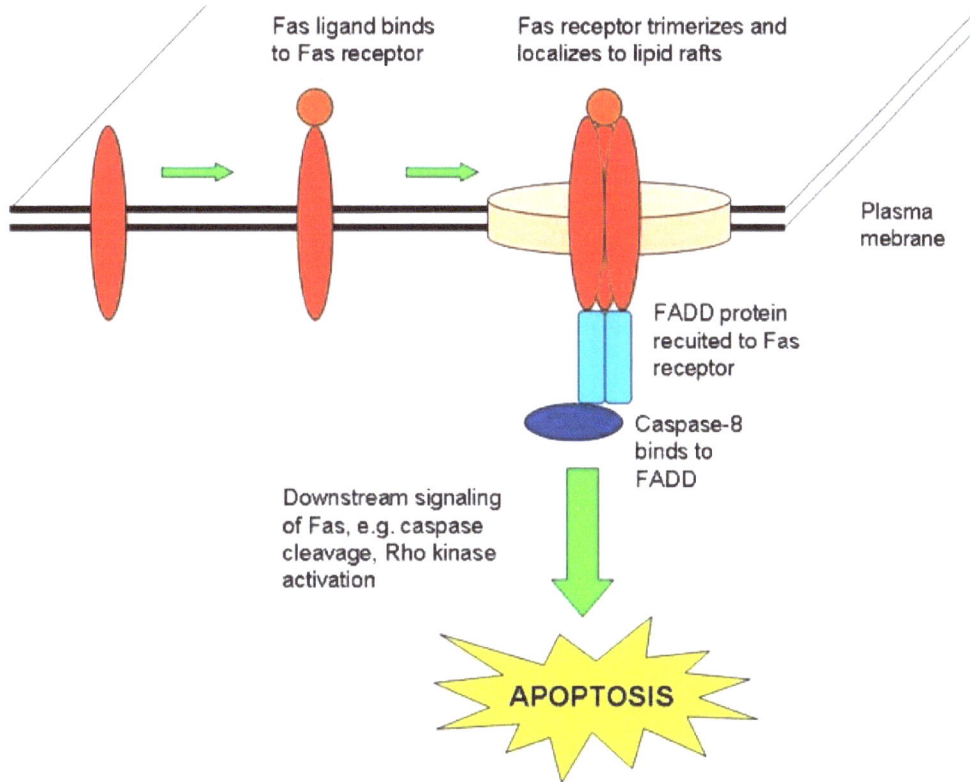

Figure 3: Signaling of the death receptor Fas.

Activation of Fas/CD95 subsequently activates downstream signalling molecules including the death-inducing signaling complex (DISC), and kinases such as Rho kinase (ROCK), eventually leading to apoptosis. The effectiveness of these downstream molecules seems to depend on the presence of the Fas/CD95 trimer in lipid rafts, for example ROCK [137], which is in itself associated with caveolin-1 [138]. Since apoptosis is an indispensible control point of cell prolife-ration and therefore tumour growth, stimulation of this programmed cell death has received attention in cancer treatment. Notably, a recent study convincingly demonstrated induction of DISC formation in lipid rafts by an anti-cancer drug [139]. In colon cancer cells, Fas/CD95 localization to rafts and thereby apoptotic signalling is impaired, resulting in tumor growth and metastasis. Pharmacological compounds that promote redistribution of Fas to lipid rafts could lead to cell death of tumorigenic colon cells be a potential therapy [140,141].

Signalling proteins related to various other cancers such as prostate cancer, breast cancer and skin cancer have also been associated with lipid rafts and caveolae, specifically caveolin-1. In prostate cancer, a positive correlation between caveolin-1 and tumor aggressiveness has been documented, however downstream signalling events of this association are largely uncharacterized. Some evidence suggests that caveolin-1 and/or lipid rafts harbour hormone receptors such as the androgen receptor [142], thereby providing a possible platform for hormonal regulation of cell growth. Interestingly, there also seems to be a connection between caveolin-1 and fatty acid synthase expression [143], which putatively could be a means of regulating cell membrane lipid composition.

NUTRITIONAL MODULATION OF MEMBRANE RAFTS BY DIETARY LIPIDS

Nutritional modification of membrane raft structure and function is a growing field of study. Studies have shown that dietary lipids can be effective in modifying the lipid environment of membrane rafts including lipid rafts and caveolae leading to changes in downstream signalling and cellular function. The liquid ordered environment of membrane rafts is dynamic and can be readily modified by chemical agents such as the sequestering of cholesterol by cyclodextrin. In addition, simply changing the dietary composition of lipids can readily alter the lipid composition and consequently the function of membrane rafts. Evidence to date suggests that specific dietary fatty acids and gangliosides can modulate membrane raft mediated signalling and function.

Due to the unique chemical (cis/trans bonds), positional (position of double bonds) and structural configuration of specific fatty acids, they can markedly affect the biophysical properties of cellular membranes. A great deal of interest has focused on the modulatory effects of polyun-saturated fatty acids (PUFA) of the n-6 and n-3 families. The two major types of n-6 PUFA include linoleic acid (LA, 18:2n-6) and arachidonic acid (AA, 20:4n6) which are found in plant and animal fats. There are 3 major types of n-3 PUFA including alpha-linolenic acid (ALA, 18:3n3) from plant oils and eicosapentaenoic acid (EPA, 20:5n3) and docosahexaenoic acid (DHA, 22:6n3) which are found together in marine oils.

Supplementation of rodents with dietary fish oil or in cell culture have been shown to modulate signal transduction pathways in mouse colonocytes [24], MDA-MB-231 breast cancer cells [144], and mouse immune cells [145]. Similar effects have also been observed with AA [146]. There have also been reports that trans fatty acids can affect membrane rafts. Conjugated linoleic acids (CLA), refers to a group of cis and trans isomers of the n-6 PUFA, linoleic acid, found commonly in ruminant fats. CLA are reported to have anticancer properties and were have been shown to incorporate and modify the fatty acid composition of caveolae of MCF-7 breast cancer cells [147]. In contrast, hydrogenated monounsaturated trans fatty acids (TFA) from vegetable oils which are associated with negative health outcomes, particularly cardiovascular disease have been shown to facilitate the formation of rafts in model membrane systems [148].

The unsaturated characteristic of PUFA gives them a high degree of flexibility which is presumably incompatible with the liquid ordered environment of membrane rafts. Therefore, their direct incorporation into membrane rafts can be viewed as disruptive or alternatively, they may have a stabilizing role. Typically, PUFA occupy the Sn2 position of phospholipids while the Sn1 position is occupied by a saturated fatty acids, therefore it has been hypothesized that if the saturated fatty acid faces inwards towards the membrane raft and the PUFA is orientated to the bulk plasma membrane that this could act as a shell to help maintain the integrity of the membrane raft [149]. In contrast, trans containing fatty acids are more similar in structure to saturated fatty acids which are preferentially enriched in membrane rafts, therefore may enhance membrane raft structure. However, the incorporation of the polyunsaturated CLA and monounsaturated TFA appear to have opposing biological effects.

In addition to dietary fatty acids, there is increasing evidence that dietary gangliosides can also affect membrane raft structure and function. Gangliosides are glycophingolipids containing one or more sialic acids, which are found in the milk fat of mammals. Gangliosides have been shown to have a beneficial effect on the function of intestinal, immune and neuronal tissues. It has been shown that dietary gangliosides can modify membrane raft structure primarily by reducing cholesterol content [150, 151] in gut epithelium.

Overall, this growing body of research demonstrates a novel modality by which dietary lipids modulates cell signalling in many cell types including normal and disease models, and therefore provides evidence for the potential to modulate disease risk. Further research is warranted to elucidate the effects of various other types of dietary lipids on membrane rafts. Also, there is need to substantiate the experimental work in humans.

CONCLUSION

Specific lipid membrane domains rich in sphingolipids and cholesterol provide unique signalling platforms in the plasma membrane. These domains harbour various signaling proteins and receptors and thereby have been linked to signalling pathways in both health and disease. These include prevalent diseases such as Alzheimer's disease,

Parkinson's disease and cardiovascular disease. In analyzing the causes and effects of these conditions, it is important to distinguish rafts in terms of their lipid and protein composition. The study of membrane rafts has elucidated the role of the plasma membrane as a signalling organelle and can putatively aid in explaining systems for disease progression and cure.

REFERENCES

[1]   Singer, S. J. & Nicolson, G.L. The fluid mosaic model of the structure of cell membranes. Science 175, 720-31 (1972).

[2]   Engelman, D. M. Membranes are more mosaic than fluid. Nature 438, 578-80 (2005).

[3]   van Meer, G. et al. Transbilayer distribution and mobility of phosphatidylcholine in intact erythrocyte membranes. A study with phosphatidylcholine exchange protein. Eur J Biochem 103, 283-8 (1980).

[4]   Kahlenberg, A. et al. Evidence for an asymmetric distribution of phospholipids in the human erythrocyte membrane. Can J Biochem 52, 803-6 (1974).

[5]   Verkleij, A. J. et al. The asymmetric distribution of phospholipids in the human red cell membrane. A combined study using phospholipases and freeze-etch electron microscopy. Biochim Biophys Acta 323, 178-93 (1973).

[6]   van Meer, G. Lipid traffic in animal cells. Annu Rev Cell Biol 5, 247-75 (1989).

[7]   van Meer, G. et al. Sorting of sphingolipids in epithelial (Madin-Darby canine kidney) cells. J Cell Biol 105, 1623-35 (1987).

[8]   Yu, J. et al. Selective solubilization of proteins and phospholipids from red blood cell membranes by nonionic detergents. J Supramol Struct 1, 233-48 (1973).

[9]   Brown, D. A. & Rose, J. K. Sorting of GPI-anchored proteins to glycolipid-enriched membrane subdomains during transport to the apical cell surface. Cell 68, 533-44 (1992).

[10]   Zurzolo, C. et al. Glycosphingolipid clusters and the sorting of GPI-anchored proteins in epithelial cells. Braz J Med Biol Res 27, 317-22 (1994).

[11]   Varma, R. & Mayer, S.. GPI-anchored proteins are organized in submicron domains at the cell surface. Nature 394, 798-801 (1998).

[12]   Pike, L. J. Rafts defined: a report on the Keystone Symposium on Lipid Rafts and Cell Function. J Lipid Res 47, 1597-8 (2006).

[13]   Munro, S. Lipid rafts: elusive or illusive? Cell 115, 377-88 (2003).

[14]   Nichols, B. Cell biology: without a raft. Nature 436, 638-9 (2005).

[15]   Heerklotz, H. Triton promotes domain formation in lipid raft mixtures. Biophys J 83, 2693-701 (2002).

[16]   Lichtenberg, D. et al. Detergent-resistant membranes should not be identified with membrane rafts. Trends Biochem Sci 30, 430-6 (2005).

[17]   Dietrich, C. et al. Partitioning of Thy-1, GM1, and cross-linked phospholipid analogs into lipid rafts reconstituted in supported model membrane monolayers. Proc Natl Acad Sci U.S.A. 98, 10642-7 (2001).

[18]   Samsonov, A. V. et al. Characterization of cholesterol-sphingomyelin domains and their dynamics in bilayer membranes. Biophys J 81, 1486-500 (2001).

[19]   Parker, A. et al. Lateral distribution of cholesterol in dioleoylphosphatidylcholine lipid bilayers: cholesterol-phospholipid interactions at high cholesterol limit. Biophys J 86, 1532-44 (2004).

[20]   Scherfeld, D. et al. Lipid dynamics and domain formation in model membranes composed of ternary mixtures of unsaturated and saturated phosphatidylcholines and cholesterol. Biophys J 85, 3758-68 (2003).

[21]   Pitman, M. C. et al. Molecular-level organization of saturated and polyunsaturated fatty acids in a phosphatidylcholine bilayer containing cholesterol. Biochemistry 43, 15318-28 (2004).

[22]   Li, X. M. et al. Sterol structure and sphingomyelin acyl chain length modulate lateral packing elasticity and detergent solubility in model membranes. Biophy J 85, 3788-801 (2003).

[23]   Ma, D. W. et al. n-3 PUFA and membrane microdomains: a new frontier in bioactive lipid research. J Nutr Biochem 15, 700-6 (2004).

[24]   Ma, D. W. et al. n-3 PUFA alter caveolae lipid composition and resident protein localization in mouse colon. FASEB J 18, 1040-2 (2004).

[25]   Yaqoob, P. The Nutritional Significance of Lipids Rafts. Annu Rev Nutr (2009).

[26]   Simons, K. & Ikonen, E. Functional rafts in cell membranes. Nature 387, 569-72 (1997).

[27] Filippov, A., Oradd, G. & Lindblom, G. Sphingomyelin structure influences the lateral diffusion and raft formation in lipid bilayers. Biophys J 15, 2086-92 (2006).

[28] London, E. & Brown, D. A. Insolubility of lipids in triton X-100: physical origin and relationship to sphingolipid/cholesterol membrane domains (rafts). Biochim Biophys Acta 1508, 182-95 (2000).

[29] Stuven, E. et al. Intra-Golgi protein transport depends on a cholesterol balance in the lipid membrane. J Biol Chem 278, 53112-22 (2003).

[30] Zheng, Y. Z. et al. Mitochondria do not contain lipid rafts, and lipid rafts do not contain mitochondrial proteins. J Lipid Res 50, 988-98 (2009).

[31] Gilch, S. et al. The prion protein requires cholesterol for cell surface localization. Mol Cell Neurosci 31, 346-53 (2006).

[32] Ostrom, R. S. et al. Nitric oxide inhibition of adenylyl cyclase type 6 activity is dependent upon lipid rafts and caveolin signaling complexes. J Biol Chem 279, 19846-53 (2004).

[33] Vial, C. & Evans, R. J. Disruption of lipid rafts inhibits P2X1 receptor-mediated currents and arterial vasoconstriction. J Biol Chem 280, 30705-11 (2005).

[34] Maguy, A. et al. Involvement of lipid rafts and caveolae in cardiac ion channel function. Cardiovasc Res 69, 798-807 (2006).

[35] Ishitsuka, R. et al. Imaging lipid rafts. J Biochem (Tokyo) 137, 249-54 (2005).

[36] Nosjean, O. et al. Mammalian GPI proteins: sorting, membrane residence and functions. Biochim Biophys Acta 1331, 153-86 (1997).

[37] Maeda, Y. et al. PIG-M transfers the first mannose to glycosylphosphatidylinositol on the lumenal side of the ER. EMBO J 20, 250-61 (2001).

[38] Brown, D. A. & London, E. Structure and origin of ordered lipid domains in biological membranes. J Membr Biol 164, 103-14 (1998).

[39] Bagnat, M. et al. Lipid rafts function in biosynthetic delivery of proteins to the cell surface in yeast. Proc Natl Acad Sci USA 97, 3254-9 (2000).

[40] Scheiffele, P. et al. Interaction of influenza virus haemagglutinin with sphingolipid-cholesterol membrane domains via its transmembrane domain. EMBO J 16, 5501-8 (1997).

[41] Murata, M. et al. VIP21/caveolin is a cholesterol-binding protein. Proc Natl Acad Sci USA 92, 10339-43 (1995).

[42] Marquez, D. C. et al. Estrogen receptors in membrane lipid rafts and signal transduction in breast cancer. Mol Cell Endocrinol 246, 91-100 (2006).

[43] Pike, L. J. Growth factor receptors, lipid rafts and caveolae: an evolving story. Biochim Biophys Acta 1746, 260-73 (2005).

[44] Pani, B. & Singh, B. B. Lipid rafts/caveolae as microdomains of calcium signaling. Cell Calcium 45, 625-33 (2009).

[45] Park, Y. & Kim, K. T. Dominant role of lipid rafts L-type calcium channel in activity-dependent potentiation of large dense-core vesicle exocytosis. J Neurochem 110, 520-29 (2009).

[46] El-Yazbi, A. F. et al. Calcium extrusion by plasma membrane calcium pump is impaired in caveolin-1 knockout mouse small intestine. Eur J Pharmacol 591, 80-7 (2008).

[47] Liu, L. & Askari, A. Beta-subunit of cardiac Na+-K+-ATPase dictates the concentration of the functional enzyme in caveolae. Am J Physiol Cell Physiol 291, C569-78 (2006).

[48] Ruppelt, A. et al. Inhibition of T cell activation by cyclic adenosine 5'-monophosphate requires lipid raft targeting of protein kinase A type I by the A-kinase anchoring protein ezrin. J Immunol 179, 5159-68 (2007).

[49] Okamoto, T. et al. Caveolins, a family of scaffolding proteins for organizing "preassembled signaling complexes" at the plasma membrane. J Biol Chem 273, 5419-22 (1998).

[50] Parton, R. G. et al. Biogenesis of caveolae: a structural model for caveolin-induced domain formation. J Cell Sci 119, 787-96 (2006).

[51] Stahl, P. & Schwartz, A. L. Receptor-mediated endocytosis. J Clint Invest **77**, 657-62 (1986).

[52] Nichols, B. J. & Lippincott-Schwartz, J. Endocytosis without clathrin coats. Trends Cell Biol 11, 406-12 (2001).

[53] Puri, V. et al. Clathrin-dependent and -independent internalization of plasma membrane sphingolipids initiates two Golgi targeting pathways. J Cell Biol 154, 535-47 (2001).

[54] Lamaze, C. et al. Interleukin 2 receptors and detergent-resistant membrane domains define a clathrin-independent endocytic pathway. Mol Cell **7**, 661-71 (2001).

[55]   Oh, P. et al. Dynamin at the neck of caveolae mediates their budding to form transport vesicles by GTP-driven fission from the plasma membrane of endothelium. J Cell Bio 141, 101-14 (1998).

[56]   Glebov, O. O. et al. Flotillin-1 defines a clathrin-independent endocytic pathway in mammalian cells. Nat Cell Biol 8, 46-54 (2006).

[57]   Mundy, D. I. et al. Dual control of caveolar membrane traffic by microtubules and the actin cytoskeleton. J Cell Sci 115, 4327-39 (2002).

[58]   Norkin, L. C. et al. Caveolar endocytosis of simian virus 40 is followed by brefeldin A-sensitive transport to the endoplasmic reticulum, where the virus disassembles. J Virol 76, 5156-66 (2002).

[59]   Nunes-Correia, I. et al. Caveolae as an additional route for influenza virus endocytosis in MDCK cells. Cell Mol Biol Lett 9, 47-60 (2004).

[60]   Hu, L. et al. Signal transduction events involved in human epithelial cell invasion by Campylobacter jejuni 81-176. Microb Pathog 40, 91-100 (2006).

[61]   Sukumaran, S. K. et al. Escherichia coli K1 internalization via caveolae requires caveolin-1 and protein kinase Calpha interaction in human brain microvascular endothelial cells. J Biol Chem 277, 50716-24 (2002).

[62]   Fra, A. M. et al. De novo formation of caveolae in lymphocytes by expression of VIP21-caveolin. Proc Natl Acad Sci USA 92, 8655-9 (1995).

[63]   Kurzchalia, T. V. & Parton, R. G. Membrane microdomains and caveolae. *Curr.* Opin Cell Biol 11, 424-31 (1999).

[64]   Le, P. U. et al. Caveolin-1 is a negative regulator of caveolae-mediated endocytosis to the endoplasmic reticulum. J Biol Chem 277, 3371-9 (2002).

[65]   Pelkmans, L. et al. Caveolar endocytosis of simian virus 40 reveals a new two-step vesicular-transport pathway to the ER. Nat Cell Biol 3, 473-83 (2001).

[66]   Lundmark, R. et al. The GTPase-activating protein GRAF1 regulates the CLIC/GEEC endocytic pathway. Curr Biol 18, 1802-8 (2008).

[67]   Bhagatji, P. et al. Steric and not structure-specific factors dictate the endocytic mechanism of glycosylphosphatidylinositol-anchored proteins. J Cell Biol 186, 615-28 (2009).

[68]   Gonzalez, E. et al. Small interfering RNA-mediated down-regulation of caveolin-1 differentially modulates signaling pathways in endothelial cells. J Biol Chem 279, 40659-69 (2004).

[69]   Zuo, L. et al. Caveolin-1 is essential for activation of Rac1 and NAD(P)H oxidase after angiotensin II type 1 receptor stimulation in vascular smooth muscle cells: role in redox signaling and vascular hypertrophy. Arterioscler Thromb Vasc Biol 25, 1824-30 (2005).

[70]   Woodman, S. E. et al. Caveolin-3 knock-out mice develop a progressive cardiomyopathy and show hyperactivation of the p42/44 MAPK cascade. J Biol Chem 277, 38988-97 (2002).

[71]   Pohl, J. et al. Long-chain fatty acid uptake into adipocytes depends on lipid raft function. Biochemistry 43, 4179-87 (2004).

[72]   Head, B. P. et al. G-protein-coupled receptor signaling components localize in both sarcolemmal and intracellular caveolin-3-associated microdomains in adult cardiac myocytes. J Biol Chem 280, 31036-44 (2005).

[73]   Fecchi, K. et al. Spatial and temporal regulation of GLUT4 translocation by flotillin-1 and caveolin-3 in skeletal muscle cells. FASEB J 20, 705-57 (2006).

[74]   Mazzone, A. et al. Isolation and characterization of lipid microdomains from apical and basolateral plasma membranes of rat hepatocytes. Hepatology 43, 287-96 (2006).

[75]   Haeryfar, S. M. & Hoskin, D. W. Selective pharmacological inhibitors reveal differences between Thy-1- and T cell receptor-mediated signal transduction in mouse T lymphocytes. Int Immunopharmacol 1, 689-98 (2001).

[76]   Conrad, D. M. et al. Role of mitogen-activated protein kinases in Thy-1-induced T-lymphocyte activation. Cell Signal 21, 1298-307 (2009).

[77]   Krishnan, S. et al. Alterations in lipid raft composition and dynamics contribute to abnormal T cell responses in systemic lupus erythematosus. J Immunol 172, 7821-31 (2004).

[78]   Pavon, E. J. et al. Increased association of CD38 with lipid rafts in T cells from patients with systemic lupus erythematosus and in activated normal T cells. Mol Immunol 43, 1029-39 (2006).

[79]   Jury, E. C. et al. Altered lipid raft-associated signaling and ganglioside expression in T lymphocytes from patients with systemic lupus erythematosus. J Clin Invest 113, 1176-87 (2004).

[80] Mukasa, R. et al. Characterization of glycosylphosphatidylinositol (GPI)-anchored NCAM on mouse skeletal muscle cell line C2C12: the structure of the GPI glycan and release during myogenesis. Arch Biochem Biophys 318, 182-90 (1995).

[81] Niethammer, P. et al. Cosignaling of NCAM via lipid rafts and the FGF receptor is required for neuritogenesis. J Cell Biol **157**, 521-32 (2002).

[82] Barker, T. H. et al. Thrombospondin-1-induced focal adhesion disassembly in fibroblasts requires Thy-1 surface expression, lipid raft integrity, and Src activation. J Biol Chem 279, 23510-6 (2004).

[83] Prusiner, S. B. Prions. Proc Natl Acad Sci U.S.A. 95, 13363-83 (1998).

[84] Chiarini, L. B. et al. Cellular prion protein transduces neuroprotective signals. EMBO J 21, 3317-26 (2002).

[85] Solforosi, L. et al. Cross-linking cellular prion protein triggers neuronal apoptosis *in vivo* Science 303, 1514-6 (2004).

[86] Mouillet-Richard, S. et al. Signal transduction through prion protein. Science 289, 1925-8 (2000).

[87] Naslavsky, N. et al. Characterization of detergent-insoluble complexes containing the cellular prion protein and its scrapie isoform. J Biol Chem 272, 6324-31 (1997).

[88] Baron, G. S. & Caughey, B. Effect of glycosylphosphatidylinositol anchor-dependent and -independent prion protein association with model raft membranes on conversion to the protease-resistant isoform. J Biol Chem 278, 14883-92 (2003).

[89] Walmsley, A. R. et al. The N-terminal region of the prion protein ectodomain contains a lipid raft targeting determinant. J Biol Chem 278, 37241-8 (2003).

[90] Sarnataro, D. et al. PrP(C) association with lipid rafts in the early secretory pathway stabilizes its cellular conformation. Mol Biol Cell 15, 4031-42 (2004).

[91] Russelakis-Carneiro, M. et al. Prion replication alters the distribution of synaptophysin and caveolin 1 in neuronal lipid rafts. Am J Pathol 165, 1839-48 (2004).

[92] Brugger, B. et al. The membrane domains occupied by glycosylphosphatidylinositol-anchored prion protein and Thy-1 differ in lipid composition. J Biol Chem 279, 7530-6 (2004).

[93] Sarnataro, D. et al. Lipid rafts and clathrin cooperate in the internalization of PrP in epithelial FRT cells. PLoS One 4, e5829 (2009).

[94] Taylor, D. R. et al. Assigning functions to distinct regions of the N-terminus of the prion protein that are involved in its copper-stimulated, clathrin-dependent endocytosis. J Cell Sci 118, 5141-53 (2005).

[95] Simons, M. et al. Cholesterol and Alzheimer's disease: is there a link? Neurology 57, 1089-93 (2001).

[96] Ehehalt, R. et al. Amyloidogenic processing of the Alzheimer beta-amyloid precursor protein depends on lipid rafts. J Cell Biol 160, 113-23 (2003).

[97] Zha, Q. et al. GM1 ganglioside regulates the proteolysis of amyloid precursor protein. Mol Psychiatry 9, 946-52 (2004).

[98] Simons, M. et al. Cholesterol depletion inhibits the generation of beta-amyloid in hippocampal neurons. Proc Natl Acad Sci U.S.A. 95, 6460-4 (1998).

[99] Kakio, A. et al. Cholesterol-dependent formation of GM1 ganglioside-bound amyloid beta-protein, an endogenous seed for Alzheimer amyloid. J Biol Chem 276, 24985-90 (2001).

[100] Molander-Melin, M. et al. Structural membrane alterations in Alzheimer brains found to be associated with regional disease development; increased density of gangliosides GM1 and GM2 and loss of cholesterol in detergent-resistant membrane domains. J Neurochem 92, 171-82 (2005).

[101] Barbeau, A. Dopamine and dopamine metabolites in Parkinson's disease-a review. Proc Aust Assoc Neurol 5, 95-100 (1968).

[102] Tolosa, E. et al. The diagnosis of Parkinson's disease. Lancet Neurol 5, 75-86 (2006).

[103] Rippon, G. A. & Marder, K. S. Dementia in Parkinson's disease. Adv Neurol 96, 95-113 (2005).

[104] Mentis, M. J. & Delalot, D. Depression in Parkinson's disease. Adv Neurol 96, 26-41 (2005).

[105] Richard, I. H. Anxiety disorders in Parkinson's disease. Adv Neurol 96, 42-55 (2005).

[106] Polymeropoulos, M. H. et al. Mutation in the alpha-synuclein gene identified in families with Parkinson's disease. Science 276, 2045-7 (1997).

[107] Liu, C. W. et al. A precipitating role for truncated alpha-synuclein and the proteasome in alpha-synuclein aggregation: implications for pathogenesis of Parkinson disease. J Biol Chem 280, 22670-8 (2005).

[108] Eriksen, J. L. et al. Molecular pathogenesis of Parkinson disease. Arch Neurol 62, 353-7 (2005).

[109] Hashimoto, M. et al. Alpha-synuclein up-regulates expression of caveolin-1 and down-regulates extracellular signal-regulated kinase activity in B103 neuroblastoma cells: role in the pathogenesis of Parkinson's disease. J Neurochem 85, 1468-79 (2003).

[110] Fortin, D. L. et al. Lipid rafts mediate the synaptic localization of alpha-synuclein. J Neurosci 24, 6715-23 (2004).

[111] Kubo, S. et al. A combinatorial code for the interaction of alpha-synuclein with membranes. J Biol Chem 280, 31664-72 (2005).

[112] Jacobowitz, D. M. & Kallarakal, A. T. Flotillin-1 in the substantia nigra of the Parkinson brain and a predominant localization in catecholaminergic nerves in the rat brain. Neurotox Res 6, 245-57 (2004).

[113] Rosse, W. F. & Ware, R. E. The molecular basis of paroxysmal nocturnal hemoglobinuria. Blood 86, 3277-86 (1995).

[114] Morgan, B. P. et al. Cross-linking of CD59 and of other glycosyl phosphatidylinositol-anchored molecules on neutrophils triggers cell activation via tyrosine kinase. Eur J Immunol 23, 2841-50 (1993).

[115] Murray, E. W. & Robbins, S. M. Antibody cross-linking of the glycosylphosphatidylinositol-linked protein CD59 on hematopoietic cells induces signaling pathways resembling activation by complement. J Biol Chem 273, 25279-84 (1998).

[116] Almeida, A. M. et al. Hypomorphic promoter mutation in PIGM causes inherited glycosylphosphatidylinositol deficiency. Nat Med 12, 846-51 (2006).

[117] Song, K. S. et al. Expression of caveolin-3 in skeletal, cardiac, and smooth muscle cells. Caveolin-3 is a component of the sarcolemma and co-fractionates with dystrophin and dystrophin-associated glycoproteins. J Biol Chem 271, 15160-5 (1996).

[118] Kikuchi, T. et al. Behavior of caveolae and caveolin-3 during the development of myocyte hypertrophy. J Cardiovasc Pharmacol 45, 204-10 (2005).

[119] Jeong, K. et al. Modulation of the caveolin-3 localization to caveolae and STAT3 to mitochondria by catecholamine-induced cardiac hypertrophy in H9c2 cardiomyoblasts. Exp Mol Med 41, 226-35 (2009).

[120] Augustus, A. S. et al. Substrate uptake and metabolism are preserved in hypertrophic caveolin-3 knockout hearts. Am J Physiol Heart Circ Physiol 295, H657-66 (2008).

[121] Insel, P. A. et al. Caveolae and lipid rafts: G protein-coupled receptor signaling microdomains in cardiac myocytes. Ann N Y Acad Sci 1047, 166-72 (2005).

[122] Couet, J. et al. Interaction of a receptor tyrosine kinase, EGF-R, with caveolins. Caveolin binding negatively regulates tyrosine and serine/threonine kinase activities. J Biol Chem 272, 30429-38 (1997).

[123] Fujita, T. et al. Caveolin-3 inhibits growth signal in cardiac myoblasts in a Ca2+-dependent manner. J Cell Mol Med 10, 216-24 (2006).

[124] Koga, A. et al. Adenovirus-mediated overexpression of caveolin-3 inhibits rat cardiomyocyte hypertrophy. Hypertension 42, 213-9 (2003).

[125] Zhang, A. Y. et al. Lipid raft clustering and redox signaling platform formation in coronary arterial endothelial cells. Hypertension 47, 74-80 (2006).

[126] Ushio-Fukai, M. et al. cAbl tyrosine kinase mediates reactive oxygen species- and caveolin-dependent AT1 receptor signaling in vascular smooth muscle: role in vascular hypertrophy. Circ Res 97, 829-36 (2005).

[127] Steinberg, S. F. beta(2)-Adrenergic receptor signaling complexes in cardiomyocyte caveolae/lipid rafts. J Mol Cell Cardiol 37, 407-15 (2004).

[128] Fujita, T. et al. Accumulation of molecules involved in alpha1-adrenergic signal within caveolae: caveolin expression and the development of cardiac hypertrophy. Cardiovasc Res 51, 709-16 (2001).

[129] Feron, O. et al. Dynamic targeting of the agonist-stimulated m2 muscarinic acetylcholine receptor to caveolae in cardiac myocytes. J Biol Chem 272, 17744-8 (1997).

[130] Yarbrough, T. L. et al. Localization of cardiac sodium channels in caveolin-rich membrane domains: regulation of sodium current amplitude. Circ Res 90, 443-9 (2002).

[131] Cavalli, A. et al. Localization of sarcolemmal proteins to lipid rafts in the myocardium. Cell Calcium 42, 313-22 (2007).

[132] Der, P. et al. Role of lipid rafts in ceramide and nitric oxide signaling in the ischemic and preconditioned hearts. J Mol Cell Cardiol 40, 313-20 (2006).

[133] Hueber, A. O. et al. An essential role for membrane rafts in the initiation of Fas/CD95-triggered cell death in mouse thymocytes. EMBO Rep 3, 190-6 (2002).

[134] Mollinedo, F. & Gajate, C. Fas/CD95 death receptor and lipid rafts: new targets for apoptosis-directed cancer therapy. Drug Resist Updat 9, 51-73 (2006).

[135] Scaffidi, C. et al. Two CD95 (APO-1/Fas) signaling pathways. EMBO J 17, 1675-87 (1998).

[136] Gniadecki, R. Depletion of membrane cholesterol causes ligand-independent activation of Fas and apoptosis. Biochem Biophys Res Commun 320, 165-9 (2004).

[137] Soderstrom, T. S. et al. CD95 capping is ROCK-dependent and dispensable for apoptosis. J Cell Sci 118, 2211-23 (2005).

[138] Rashid-Doubell, F. et al. Caveolin-1 and Lipid Rafts in Confluent BeWo Trophoblasts: Evidence for Rock-1 Association with Caveolin-1. Placenta (2006).

[139] Gajate, C. et al. Involvement of raft aggregates enriched in Fas/CD95 death-inducing signaling complex in the antileukemic action of edelfosine in Jurkat cells. PLoS One 4, e5044 (2009).

[140] Lacour, S. et al. Cisplatin-induced CD95 redistribution into membrane lipid rafts of HT29 human colon cancer cells. Cancer Res 64, 3593-8 (2004).

[141] Delmas, D. et al. Resveratrol-induced apoptosis is associated with Fas redistribution in the rafts and the formation of a death-inducing signaling complex in colon cancer cells. J Biol Chem 278, 41482-90 (2003).

[142] Sun, M. et al. Activation of phosphatidylinositol 3-kinase/Akt pathway by androgen through interaction of p85alpha, androgen receptor, and Src. J Biol Chem 278, 42992-3000 (2003).

[143] Di Vizio, D. et al. Caveolin-1 is required for the upregulation of fatty acid synthase (FASN), a tumor promoter, during prostate cancer progression. Cancer Biol Ther 6, 1263-8 (2007).

[144] Schley, P. D. et al. (n-3) PUFA alter raft lipid composition and decrease epidermal growth factor receptor levels in lipid rafts of human breast cancer cells. J Nutr 137, 548-53 (2007).

[145] Fan, Y. Y. et al. Dietary (n-3) polyunsaturated fatty acids remodel mouse T-cell lipid rafts. J Nutr 133, 1913-20 (2003).

[146] Webb, Y. et al. Inhibition of protein palmitoylation, raft localization, and T cell signaling by 2-bromopalmitate and polyunsaturated fatty acids. J Biol Chem 275, 261-70 (2000).

[147] Huot, P. S.& Ma, W.L. CLA incorporates into caveolae phospholipids and reduces caveolin-1 expression in MCF-7 cells. Submitted for publication. Presented at the 98th American Oil Chemists' Society (AOCS) Annual Meeting; Quebec, Canada. May, 2007. AOCS Press 81 (2007).

[148] Bjorkbom, A. et al. Phosphatidylcholine and sphingomyelin containing an elaidoyl fatty acid can form cholesterol-rich lateral domains in bilayer membranes. Biochim Biophys Acta 1768, 1839-47 (2007).

[149] Wassall, S. R. & Stillwell, W. Polyunsaturated fatty acid-cholesterol interactions: domain formation in membranes. Biochim Biophys Acta 1788, 24-32 (2009).

[150] Park, E. J. et al. Dietary ganglioside inhibits acute inflammatory signals in intestinal mucosa and blood induced by systemic inflammation of Escherichia coli lipopolysaccharide. Shock 28, 112-7 (2007).

[151] Park, E. J. et al. Dietary ganglioside decreases cholesterol content, caveolin expression and inflammatory mediators in rat intestinal microdomains. Glycobiology 15, 935-42 (2005).

CHAPTER 2

Trans-Cellular Mobility of GPI-Anchored Proteins

Barbara Viljetić, Marija Heffer-Lauc and Gordan Lauc

Abstract; Glycosylphosphatidylinositol (GPI)-anchored proteins are comprised of a relatively large hydrophilic moiety tethered to a membrane by a relatively small lipid tail. The functional role of the anchor is not clearly established, and proposals range from cell motility to cell signaling. Together with gangliosides and cholesterol, GPI-anchored proteins form membrane microdomains called lipid rafts in the outer leaflet of the plasma membrane that are involved in the regulation and modulation of numerous cellular processes.

GPI-anchored proteins can be actively shed from the membrane of one cell and taken up by other cells by insertion of their lipid anchors into the cell membrane. There are physiological processes where trans-cellular mobility of GPI-anchored proteins occurs. This process appears to be regulated, and most probably involves catalytic activity of some proteins that still have to be identified (GPI-specific phospholipases). Tumor cells and some pathogens apparently misuse this process for their own advantage, but its real physiological functions remain to be discovered. Special attention should be given when using detergents for biochemical and immunohistochemical methods in analysis of these molecules because of probable artifacts.

INTRODUCTION

Glycosylphosphatidylinositol (GPI)-anchored proteins are widely distributed among eukaryotic organisms. They have lipid tails that tether them to the cell membrane and relatively large hydrophilic domains that protrude into the extracellular space. The evolutionary purpose of using GPI linkage for cell surface protein expression is not clear, but these unusual structures are involved in a number of different physiological processes. GPI-anchored proteins play important role in a range of biological processes and serve a variety of functions, such as catalytic, signal transduction, receptors, adhesion, and complement regulation. Recently, it has been shown that pathogens need GPIs to support their survival and invasions. Aerolysin, a toxin from Aeromonas hydrophilia, requires binding to surface GPI for pore-forming and its cell-killing activity [1]. GPI-anchored proteins also have a great role in embryogenesis - abrogation in their biosynthesis results in embryonic lethality [2-4]. Recently, it was shown that interaction of GPI-anchored EphrinA with his receptor EphA is crucial in closure of the neural tube [5].

Many reviews have been recently published on GPI-anchored proteins [6-9] and lipid rafts, and they should be consulted for more detailed consideration of these molecules and their functions. Trans-cellular mobility of GPI-anchored proteins will be explored in this chapter. Their ability to be cleaved by specific phospholipases resulting in the release of the protein from the membrane will be explored but the focus will be on their other peculiar characteristic: the ability to be released from the membrane of one cell and incorporated into the membrane of another cell. This process has been independently discovered and rediscovered in several research areas during the past decades and given different names, including shedding, release, incorporation, uptake and jumping. In this manuscript, we shall refer to it as "shedding and uptake", as originally suggested by Ladisch and colleagues [15].

SYNTHESIS AND TARGETING OF GPI-ANCHORED PROTEINS

GPI-anchored proteins are a large class of proteins of extreme diversity in structure and function, and their only common feature is that they all share glycolipid membrane anchors. First indications that proteins might be attached to the cell membrane by lipid anchors appeared in 1963 with the finding that bacterial phospholipase can release alkaline phosphatase from cells [16].The presence of inositol-containing phospholipid protein anchors was postulated by Ikezawa and colleagues in 1976 [17], but their hypothesis was not widely accepted until 1985, when a body of compositional data about Torpedo electric organ acetylcholinesterase (AChE) [18], human erythrocyte AChE [19], rat brain and thymocyte Thy-1 [20], and *Trypanosoma brucei* variant surface glycoprotein (VSG) [21,22] became available. Today, hundreds of GPI-anchored proteins are known (see examples in Table **1**) and it is estimated that approximately 0.5% of all proteins in lower and higher eukaryotes are being modified in this manner [23].

John A. Dangerfield & Christoph Metzner (Eds)

Table 1: Examples of GPI-Anchored Proteins (for a more Complete List See a Review by H. Ikezawa[6])

Enzymes	Receptors	Other proteins
Alkaline phosphatase	Plasmodium transferrin receptor	Thy-1
Acetylcholinesterase	CD14	CD24
5′-Nucleotidase	CD16	CD55 (DAF)
Alkaline phosphodiesterase I	CD48	CD58
Renal dipeptidase (MDP)	Folate-binding protein	Ly6 family (CD59, Ly6A)
Aminopeptidase P	Urokinase receptor	Carcinoembryonic antigen
NAD$^+$ glycohydrolase	CNTF receptor	Prions (PrPC, PrPSc)
Carboxypeptidase M	Nogo-66 receptor	NCAM-120fs
Carbonic anhydrase IV		Tamm-Horsfall glycoprotein
ADP-ribosyltransferase		

All GPIs share a common core structure [24]. Phosphatidylinositol is glycosidically linked through carbon-6 of the inositol ring to the reducing end of a non-acetylated glucosamine moiety. Interestingly, GPIs are one of the rare instances in nature where glucosamine is found without either an acetyl group (present in most glycoconjugates) or a sulfate moiety (present in heparin) attached to the amino-group at the carbon-2 position. Three mannosyl residues, linked α1–4, α1–6, and α1–2, respectively, are attached to the glucosamine. The terminal α1–2 linked mannose is linked to phosphoethanolamine by a phosphodiester linkage. The GPI is attached to the carboxy-terminal carboxyl group of the protein by an amide linkage to the amino group of phosphoethanolamine (Fig. **1**). This common core structure can be further modified in a way that depends on both the organism and the cell type in which it is synthesized [6].

The whole process of GPI biosynthesis is carried out in the endoplasmic reticulum [25] and nearly 20 enzymes participate in this pathway. Corresponding genes have been cloned from mammals, yeast and protozoa [26]. Once it is completed, the pre-formed anchor is transferred to a specific site upstream of the C-terminal end of the protein in the ER lumen by the action of a transamidase complex, which simultaneously cleaves off the remaining C-terminal peptide [7]. GPI-modified proteins are then transferred from ER to Golgi complex where they are subjected to further modification in the GPI moiety. Finally, they are transferred to the plasma membrane through *trans*-Golgi network as mature GPI-anchored proteins6. The C-terminal sequence of the protein thus acts as a signal for GPI attachment in specialized microdomains on the cell surface. For specific internalization on the plasma membrane, GPI-anchored proteins use a clathrin-independent pathway but it requires dynamin and flotillin, contrary to transmembrane proteins which are clathrin-dependent [27].

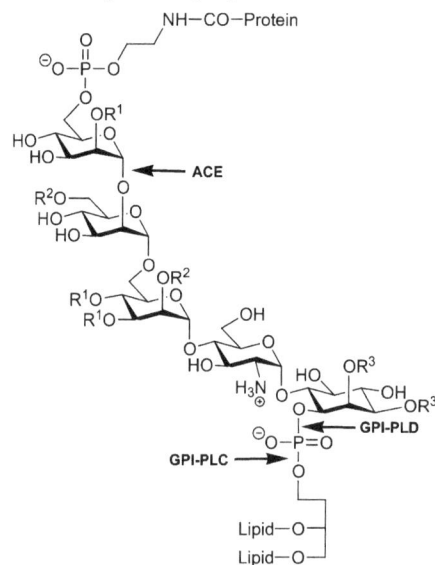

Figure 1: Structure of a GPI anchor. All characterized GPI anchors share a common core consisting of ethanolamine-PO4-6Manα1-2Manα1-6Manα1-4GlcNα1-6myo-Ino-1-PO4 lipid. Heterogeneity in GPI anchors is derived from various substitutions

of this core structure that are represented as R groups. Various glycans can be attached to R1, phosphoetanolamine is frequently found at R2, and additional fatty acids can be attached at R3. Cleavage sites of GPI phospholipase C (GPI-PLC), GPI-phospholipase D (GPI-PLD) and angiotensin-converting enzyme (ACE) are marked by arrows.

The initial step of GPI synthesis, attachment of N-acetylglucosamine to phosphatidylinositol, depends on the product of X chromosome gene termed phosphatidylinositol glycan class A (PIG-A in humans, Pig-a in mice) [27]. A deficiency in PIG-A results in a rare human disease named paroxysmal nocturnal hemoglobinuria (PNH) [27-30]. Patients with PNH have abnormal cells of various hematopoietic lineages that are defective in the biosynthesis of GPI anchored proteins. These include the complement-regulatory proteins, CD55 and CD59, whose absence results in enhanced complement-mediated lysis [31,32]. Since deficiency of GPI is embryonically lethal [2-4], all PNH patients reported to date acquired a somatic mutation in PIG-A [33]. The exact mechanism how one or a few of the large number of pluripotent hematopoietic stem cells that bear mutation in PIG-A achieve dominance in the bone marrow and the peripheral blood is not known [34], but it has been recently shown that PIG-A deficient cells have lower susceptibility to TNF-α and IFN-γ, what might contribute to their clonal dominance [35].

Although they do not apparently share common features, the presence of the anchor itself appears to confer some important behavioral attributes to proteins to which it is attached. In particular, localization to lipid raft microdomains and cleavage by endogenous and exogenous phospholipases appears to play a major role in the transduction of signals across the plasma membrane (reviewed by Sharom) [9]. Recent observation that prion protein and Thy-1 exist in separate lipid rafts, and that the composition of membrane lipids in rafts containing prion protein is different from the composition of lipids in rafts containing Thy-1 [36], suggests that interplay of lipids and GPI-linked proteins in lipid rafts is very specific and carefully regulated.

LIPID RAFTS

GPI-anchored proteins associate with membrane microdomains (lipid rafts) enriched in glycosphingolipids, cholesterol and certain types of proteins and serve as platforms for variety of cellular functions, such as vesicular trafficking and signal transduction [37,38]. Lipid rafts were first characterized by their insolubility at 4°C in the nonionic detergent Triton X-100, which has become the most widely used assay for raft resistance [37]. Some signaling proteins are also found in these complexes and that led to the hypothesis that the GPI anchor may be important in signal transduction [38]. Lack of GPI-anchored proteins alters the composition of lipid rafts and those cells with deficient GPI-anchored proteins may induce different inflammatory responses, like in, for example, PNH disorder [39].

Even though the existence and importance of lipid rafts in living cells is still being actively debated [37,40], several lines of evidence strongly support this hypothesis. For example, measurements using fluorescent folate in fluorescence resonance energy transfer (FRET) showed that GPI anchored protein are organized in cholesterol-dependent microdomains with diameters less than 70 nm in living cells [41]. Biochemical cross-linking of GPI anchored growth hormone demonstrated that GPI-anchored proteins are in proximity in rafts [42]. Antibody cross-linking was used to segregate raft proteins from non-raft proteins and it was concluded that proteins found in those interaction coalescence because of specific lipid environments [43]. Photonic force microscopy was used to determine the size of rafts in living cells [44] and electron microscopy was used to visualize clustering of rafts in IgE signalling [44]. Homo-FRET technique was used to look at GPI anchored fluorescent proteins and to determine that small fractions of GPI proteins are present in rafts [45]. Since the existence of a biochemically distinct fraction at a low temperature does not imply that the same fraction exists at a higher temperature and thus rafts may not correspond to physiological domains, Ayuyan and Cohen identified and isolated biological rafts at 37°C and neutral pH [46]. Using techniques called targeted quantum dots and single-molecule imaging, Pinaud and colleagues studied diffusion of GPI-anchored probes and revealed the existence of non-caveolar, cholesterol and GM1-rich microdomains (within these GPI-anchored probes can dynamically partition and be compartmentalized) [47].

However, it was recently shown that crosslinking of proteins inserted into the outer leaflet of the cell membrane through artificially attached lipid anchors can also induce activation of Jurkat T cell-signaling responses, indicating that at least in some cases, the formation of artificial raft-like patches on the cell membrane might be sufficient to trigger signaling events [48]. In some signaling processes, the formation of protein clusters in the membrane was reported to depend on protein–protein, and not protein–lipid interactions [49], thus although lipid rafts apparently play an important role in mediating many signal transduction processes (Table **2**), they might be only one of several similar mechanisms.

Table 2: Examples of signal transduction processes that involve lipid rafts.

•	B-cell receptor [134]
•	EGF receptor [135]
•	Endothelial NOS [136]
•	FcɛRI receptor [137]
•	Insulin receptor [138]
•	Integrins [139]
•	T-cell receptor [140]

RELEASE OF GPI-ANCHORED PROTEINS

Release by GPI-Cleavage

The only confirmed role of GPI anchoring is to attach proteins to a membrane with an anchoring device that is resistant to most extracellular proteases and lipases [50]. But there are some hypotheses about other roles and one of them suggests that the function of the GPI anchor may be to offer a site for degradation by specific endogenous phospholipases resulting in a specifically regulated release of the protein from the cell surface [50]. That has been postulated very soon after the existence of GPI-anchors was widely accepted.

The removal of the GPI lipid moiety *in vitro* was reported to cause significant alterations in enzymatic activities [51-54] and ligand binding properties also [55-57], and it is quite likely that some GPI-anchored proteins in the membrane are actually reservoirs of inactive proteins that can be activated and rapidly released by GPI cleavage.

Two types of GPI-specific phospholipases, GPI-phospholipase C (GPI-PLC) and GPI-phospholipase D (GPI-PLD) cleave GPI on different sides of the phosphodiester bond between inositol and the lipid part of the anchor (Fig. **1**). Recently, it was demonstrated that angiotensin-converting enzyme (ACE) can also specifically cleave GPI [58].

Several bacterial species secrete PI-specific type C phospholipases, including *Bacillus cereus*, *Bacillus thuringiensis*, *Staphylococcus aureus*, *Listeria monocytogenes*, and *Clostridium novii*. These enzymes are able to hydrolyze mammalian GPI anchors, and have been extensively used in the study of structure and function of GPI-linked proteins. Several parasitic protozoans, for example, *Tripanosoma brucei* and *Leishmania*, contain endogenous GPI-PLC that converts membrane-bound proteins to hydrophilic soluble forms (reviewed by Sharom) [9]. Since the first discovery of bacterial PI-PLC, endogenous mammalian GPI-PLC have been postulated to serve as important regulatory factors, reducing surface expression of GPI-anchored proteins, while simultaneously increasing the levels of soluble protein. Chan and colleagues reported that lipoprotein lipase was released from insulin treated 3T3-L1 adipocytes by cleavage of its GPI anchor and they proposed that activation of an insulin-dependent PI-PLC was responsible [59]. Alkaline phosphatase was also reported to be released in soluble form from myocytes and adipocytes upon insulin stimulation, again suggesting the action of a phospholipase C [60,61]. Park and colleagues reported that endogenous GPI-PLC releases renal dipeptidase from kidney proximal tubules *in vitro* [62] and *in vivo* [63], but mammalian GPI-PLC has yet to be identified.

Mammalian GPI-PLD was discovered in human serum by Davitz and colleagues in 1987 [64]. Despite its high concentration in mammalian serum [65] and relatively well characterized molecular biology [66] and biochemistry [67], the physiological role of GPI-PLD is not clear. In serum, GPI-PLD is associated with HDL and is apparently not active [68]. Initial reports indicated that GPI-PLD was active against GPI-anchored proteins only in the presence of detergent, and was not able to cleave the anchors of proteins in a native membrane context [69]. Over-expression experiments indicated that it is active in endoplasmic reticulum during GPI synthesis, but also in lipid rafts [70]. Lipid fluidity and packing are the most important modulators of bacterial phospholipase ability to cleave GPI anchors [71] and modulation of membrane lipids were reported to affect GPI-PLD activity *in vitro* [72], so it is quite possible that mammalian GPI-PLD also requires particular membrane composition for activity. The fact that endogenous GPI-PLD was reported to specifically release NCAM from differentiating myoblast cells [73], receptor for urokinase-type plasminogen activator from ovarian cancer cells [74] and carcinoembryonic antigen from human colon cancer cells [75] strongly supports this hypothesis. Recently, another GPI protein cleaved by GPI-PLD was found and that is Cripto-1 (CR-1). CR-1 is an oncofetal gene which plays an important role early in development and is re-expressed in cellular transformation, and is highly expressed in several types of human carcinoma [76]. Once released from tumor cells, CR-1 interacts with surrounding endothelial cells and can initiate tumor angiogenesis [77]. This release is a biological process that might serve as a tumor marker in tumor diagnostics.

The angiotensin-converting enzyme (ACE) is a well characterized zinc peptidase that regulates blood pressure by hydrolyzing bioactive peptides such as angiotensin I and bradykinin [78]. There are two ACE isoforms: a somatic form of around 150–180 kDa, which bears two catalytically active sites, and a smaller isoform (90–110 kDa) found in the testes, which contains a single active site [79]. Kondoh and colleagues recently reported that testicular ACE can specifically release GPI-anchored proteins from the cell membrane [58]. Even when the peptidase activity is abolished by either mutation or inactivation, the enzyme could still cleave GPI-anchored proteins and restore fertility to ACE-deficient sperm. This activity is not protein-specific because it cleaves a variety of GPI-anchored proteins, and its cleavage site is located between the second and the third residue of the conserved mannose core (Fig. **1**). GPI-anchor-releasing activity of ACE requires removal of cholesterol from cell membranes, and similarly to GPI-PLD which is also widely present, but mostly inactive, ACE apparently also requires a particular form of substrate presentation on the membrane for activity. Recently, a new function of ACE was found. It was shown that ACE cleaves membrane-bound carboxypeptidase M (CPM) [80], which is distributed in a wide range of tissues and cells as membrane-anchored but insoluble form also in body fluids, such as urine, seminal plasma, and amniotic fluid. CPM is a GPI-anchored protein and, together with ACE, it is a part of the membrane-bound metallopeptidase family.

There are many enzymes that play a role in cleavage and release of GPI-anchored proteins from the cell membrane even though they do not cleave the GPI-anchor itself but, rather they cleave peptides at different sites. Some of those proteases are members of the disintegrin and metalloproteinases family (ADAMs) [81] and they are involved in the release of ectodomain of prion protein. It was shown that ADAM9 and ADAM10 cleave peptides very close to the GPI-anchor of prion protein [81] and provide one of the mechanisms of regulating levels of that protein at the cell surface.

Release of Intact GPI-Anchored Proteins from the Cell Membrane

In addition to release by enzymatic cleavage, GPI-linked proteins can be released from the cell membrane with their GPI anchors intact. This release can be in the form of membrane vesicles (exosomes), but also as small aggregates that contain some membrane lipids in addition to GPI-linked proteins [82]. Exosomes are small (50–200 nm) membrane vesicles first described in studies of reticulocyte maturation about 20 years ago [83,84] that were subsequently demonstrated to be released from various cell types [85-90]. Exosomes were initially thought to correspond to internal vesicles of multivesicular bodies that are being released in the extracellular space upon their fusion with the cell membrane, but this is apparently only one way how exosomes can be formed since glycolipids and GPI-anchored proteins already embedded in the outer leaflet of the cell membrane can also be efficiently secreted in the form of exosomes [91]. Various GPI-linked proteins, including the prion protein [92] are being actively secreted in exosomes.

This process can be quite extensive as demonstrated by reticulocytes that release approximately 50% of acetylcholinesterase in exosomes during differentiation into erythrocytes [93]. Similar vesicles named prostasomes exist in seminal plasma where they assist sperm function [94]. GPI-anchored CD59, CD55 and CD52 were found on prostasomes [95], but also in a form of small aggregates in seminal plasma [82]. While prostasomes bind to target cells and are later internalized, the kinetics of transfer of GPI-anchored molecules from aggregates into cells is consistent with direct incorporation into cell membranes [82].

Shedding and Uptake of GPI-Anchored Proteins

The phenomenon of shedding and uptake of a GPI-linked protein was reported even before GPI-anchors were discovered. While investigating phospholipid exchange between cells and artificial vesicles, Bouma and colleagues showed that acetylcholinesterase and some other erythrocyte proteins were transferred from erythrocytes to the vesicles and that this process was reversible [96]. The rate, direction, and extent of such intermembrane transfers were found to depend on the relative lipid composition and fluidity of the donor and acceptor membranes [97]. Also, it was shown that purified human DAF was inserted onto sheep erythrocytes [98]. Exogenously added DAF was freely mobile on the sheep cell surface and was able to function normally as shown by its inhibition of convertase complexes. Since then, numerous GPI anchored proteins have been incorporated into a variety of different cell types and these exogenously added GPI anchored proteins retained the same characteristics and functions as endogenously expressed GPI anchored proteins [98].

Contrary to the release of GPI-anchored proteins by phospholipases C and D that removes GPI and yields soluble protein, shedding releases proteins with intact GPI that are still able to insert into membranes of other cells. Mechanism by which transfer occur is unknown, but the lipid moiety must be intact for cell membrane insertion [98]. GPI-anchor of prion protein (PrP) plays a potential role in prion disease pathogenesis [99]. Transgenic mice expressing anchorless (secreted) version of PrP not present on the cell surface were infected with pathogenic scrapie form of PrP, but they never developed clinical prion disease [99]. Experiment with cells expressing anchorless PrP showed the same effect, those cells were resistant to scrapie infection [100]. If PrP transfer between cells, lack of GPI anchor on PrP would prevent that process and could explain why cells expressing GPI-anchorless PrP were unable to sustain a scrapie infection over time.

Cell-to-cell transfer of GPI-anchored protein has been reported in a variety of *in vitro* and also *in vivo* systems. CD59 was transferred from seminal plasma to erythrocytes and other cells [95], as well as from erythrocytes to endothelial cells in mice made transgenic for this GPI-anchored protein [101]. Thy-1 was transferred between cells in chimeric murine embryoid bodies composed of normal and PIG-A "knock-out" cells [102]. Erythrocyte studies on infected patients with trypanosomias found that their cell contained membrane-bound trypanosomal variant surface glycoprotein (VSG) [103]. High-density lipoproteins (HDL) may act as carriers of CD59 and are capable of transferring this protein to erythrocytes [104]. Transfer between membranes can occur without actual membrane fusion [82] and GPI-anchored proteins are apparently transferred through vesicles or liposomes released from the donor cell [105].

GPI-anchored molecules are clustered in lipid rafts and they actively take part in membrane vesicle formation, resulting in vesicles enriched in GPI-anchored proteins [105]. Storage of erythrocytes results in loss of both CD55 and CD59 from the erythrocyte membrane [106] and creation of erythrocyte microvesicles that are enriched in GPI linked proteins including CD55 and CD59 [107]. When erythrocytes from PNH patients that were deficient in GPI-anchored proteins were incubated with HDL preparations or erythrocyte microvesicles from normal blood donors, there was significant transfer of CD55 and CD59 to the cell surface. Pretreatment of microvesicles and HDL with phosphatidylinositol-specific phospholipase C abrogated protein transfer to deficient cells, indicating that increased cell-associated CD55 and CD59 levels were related to the insertion of an intact GPI moiety, rather than to simple adhesion [108]. In a recent elegant experiment, Sloand and colleagues confirmed the ability of GPI-linked proteins to transfer between cells *in vivo* [109]. PNH patients of group A1 blood type were given transfusions of compatible, washed group O blood. Patient's group A1 cells were distinguished from the transfused group O cells by staining with a *Dolichos biflorus* lectin that specifically binds to group A1 erythrocytes. Significant transfer of GPI-linked proteins from donor cells to patient's erythrocytes could be demonstrated as early as 1 day following transfusion and persisted for several days.

GPI-linked proteins transferred from cell to cell appear to be stable and biologically functional [82,98,101,110,111]. For example, transfer of CD55 and CD59 to erythrocytes confers resistance to complement-mediated lysis [108]. For effective transfer to occur, both the GPI anchor and the protein moiety must be intact [112]. Transferred molecules are inserted into the outer leaflet of the plasma membrane by lipid chains on the GPI moiety and soluble CD59 (that lacks GPI anchor) was found to have only 1/200th the ability of GPI-linked CD59 to inactivate complement [113]. Incubation of rat Thy-1 antigen with murine lymphocytes showed that the rat protein could incorporate into murine cells and that after the membrane uptake the exogenous protein migrated with the same lateral mobility as endogenous murine Thy-1 protein [114]. Similarly, incorporation of *Trypanosoma brucei* variant surface glycoproteins (VSG) into baby hamster kidney cells showed that the inserted VSG exhibited lateral mobility equivalent to that of endogenous VSG in *T. Brucei* [115]. Interestingly neither Thy-1 inserted into lymphocytes [114] nor CD59 incorporated into neutrophils [116] supported transmembrane signaling immediately following transfer. However, CD59 incorporated into U937 monocytic cells and allowed to equilibrate for 2 hr at 37°C showed redistribution into lipid rafts and signaled intracellular Ca2+ fluxes [117]. Therefore, exogenously introduced GPI-anchored molecules appear to become functional within the target cell membrane once they have acquired a distribution similar to that of endogenous GPI-anchored proteins, but this process is slow and can take even more than 24 hr [112,118]. Metzner and colleagues modified retroviral and lentiviral vector lipid envelope with exogenously added GPI-anchored proteins and viruses remained infectious after [119]. That could suggest that virions might be capable of uptake GPI-anchored proteins *in vivo*, especially complement proteins (CD55 and CD59) which could efficiently protect viral particles from the host immune system [119].

GPI-linked proteins were reported not to transfer spontaneously from erythrocytes to liposomes, and it was suggested that *in vivo* GPI-linked membrane proteins do not spontaneously transfer between cell membranes and that they would need some catalyst [105]. This hypothesis is also supported by the observation that CD4 engineered to acquire a GPI anchor can be efficiently transferred between cell membranes in one type of cells [120], while another cell line expressing CD4-GPI fusion protein failed to release it in any form [121]. However, the identity of a potential protein catalyst for GPI shedding and uptake is not known.

What is the Physiological Function for Shedding and Uptake?

Tumor cells use shedding and uptake to evade destruction by immune cells [15,122,123], and retroviruses exploit shedding for spreading to other cells [125], but these extensively studied mechanisms are actually only examples of a misuse of shedding and uptake, and the real reason why this process developed in the course of evolution still has to be discovered. One reported function of shedding and uptake is the transfer of GPI-linked proteins and gangliosides from prostasomes and GPI-lipid aggregates released by prostate epithelium to spermatozoa [82]. Since spermatozoa do not synthesize proteins, shedding and uptake here represent an important target cell oriented mechanism by which spermatozoa can acquire new proteins and alter their antigenicity, resistance to immune attack, or other surface properties.

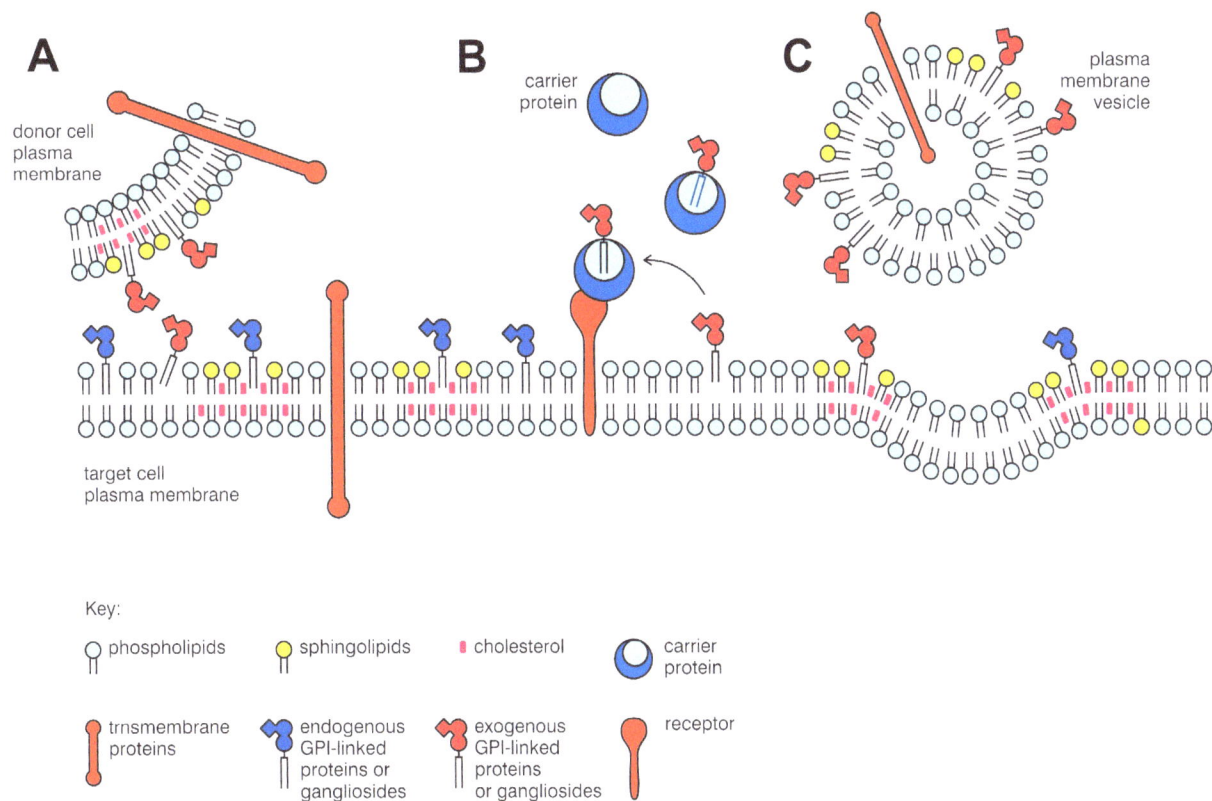

Figure 2: Shedding and uptake. GPI-linked proteins can be transferred from cell to cell either directly (A), with help of specific carrier proteins (B), or through small vesicles or micelles (C). The identities of specific proteins that catalyze shedding and uptake on the cell membrane are not known, but experimental data strongly support their existence.

Another rather probable function of shedding and uptake is the modulation of lipid rafts and signal transduction. Together with GPI-anchored proteins, gangliosides are located in lipid rafts. Even though they arise from different biochemical pathways they both have lipid anchors that tether them to the outer leaflet of cell membrane and it also allows them to be released from the membrane of one cell and inserted into the membrane of another cell in a regulated manner [125]. Since they are both localized in lipid rafts, they could have impact on each other. Exogenous administration of gangliosides affects membrane distribution of GPI-anchored proteins in lipid rafts [126,127]. Both GPI-PLD and ACE were reported to require some kind of specific membrane environment to become active, and it is tempting to speculate that modification of lipid rafts by removal or addition of specific gangliosides might create favorable conditions for activity of these enzymes and consequential release of GPI-anchored proteins. In addition to its role in the modulation of lipid rafts, shedding and uptake of gangliosides and GPI-linked proteins might be involved in some other processes.

Effect of Shedding and Uptake on Currently Used Analytic Methods

The ability of GPI-anchored proteins to move between cells in physiological conditions enables them to perform some important physiological roles, but it can also affect their behavior in various assay systems *in vitro*. Immunohistochemistry is a very important tool that enables precise localization of various types of biological molecules and structures. However, this method is prone to serious artifacts, and significant care is needed to avoid false interpretation of experimental data [128].

Cold Triton X-100 has become the most widely used assay for lipid raft isolation [37]. In addition in using in lipid raft isolation, Triton X-100 is by far the most popular detergent for improving antibody penetration for immunohistochemistry [129]. Detergents enhance the effect of shedding and uptake and their use significantly disturb the analysis of GPI-anchored proteins. Recently we reported that presence of Triton X-100 [130] (Fig. **3**) and nine other detergents [131] during immunostaining procedures caused significant redistribution of GPI-anchored proteins and gangliosides in brain sections.

Many immunostaining procedures include addition of different detergents, either to aid detection of some proteins, or to reduce background staining. However, even when all steps in the procedure are being performed at 4°C, the inclusion of even small amount of detergents in the immunostaining buffers causes significant redistribution of GPI-linked proteins from one brain region into another. In addition, tissue sections cannot be stored for a long time before immunostaining, nor can they be incubated at 37°C. Even in detergent-free solutions kept at 4°C, GPI-anchored proteins and gangliosides, which show the same effect, are being lost from the tissues during storage. The fact that immunostaining of non-fixed cells results in very little or no staining [132], is in accordance with the hypothesis that attachment of antibodies to GPI-linked proteins and gangliosides can simply pull them out of the membrane, in a kind of an *in vitro* shedding process enhanced by addition of antibodies. Fixation apparently creates some kind of mesh on the membrane which makes this more complicated.

****

PROTOCOL 2

Immunohistochemical analysis of Thy-1 GPI-anchored protein:

1.   *Collect 35-μm sections into 0.01 M phosphate buffered saline (PBS) or put 10-μm sections directly on precoated slides.*

2.   *Perform all steps of immunostaining sections at 4°C on orbital shaker.*

3.   *Pretreat sections in 0.6% hydrogen peroxide in PBS to block endogenous peroxidases.*

4.   *Block sections for 2 hr in PBS containing 1% bovine serum albumin plus 5% goat serum (Invitrogen, Carlsbad, CA, USA).*

5.   *Use blocking solution also for diluting antibodies.*

6.   *Dilute primary antibody - anti-Thy-1 (BD Biosciences, San Diego, CA) 1:2000 and incubate overnight in blocking solution.*

7.   *Wash 3 times in PBS for 15 min.*

8. *Use secondary antibody biotin-SP-AffiniPure Goat Anti-Mouse IgG (H+L) (Jackson Immunoresearch Labs., West Grove, PA, U.S.A.) in concentration of 1 µg/ml for 4 hr.*

9. *Wash 3 times in PBS for 15 min.*

10. *Follow with Vector Elite peroxidase kit (Vector Laboratories, Burlingame, CA, USA) for 2 hr.*

11. *Wash 3 times in PBS for 15 min.*

12. *Develop with SIGMAFAST™ DAB with Metal Enhancer (Sigma-Aldrich, St. Louis, MO, U.S.A).*

13. *Wash 3 times in PBS for 15 min.*

14. *Mount sections with VectaMount (Vector Laboratories, Burlingame, CA) and cover with cover glass.*

****

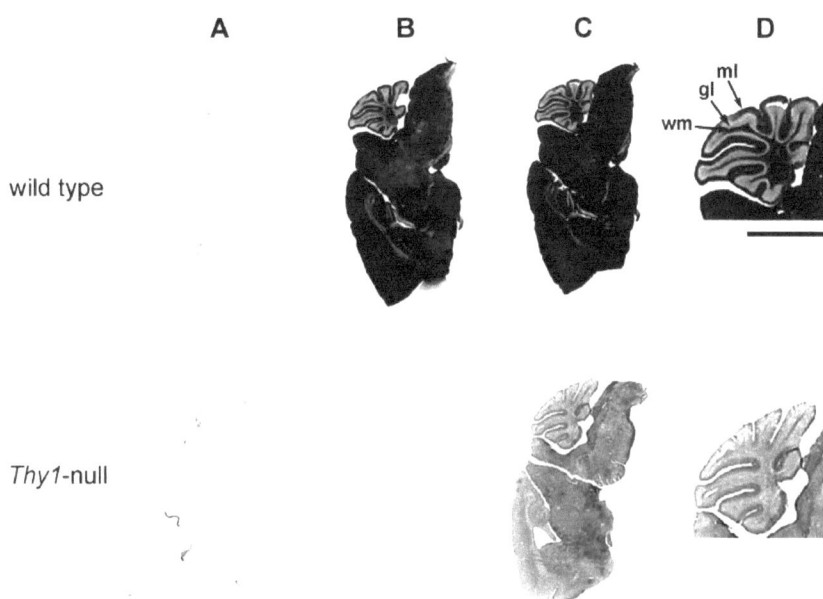

Figure 3: Effects of Triton X-100 on the histological distribution of Thy-1. Brain sections from wild-type and Thy1-null mice were incubated separately or together (as indicated) for 4 hr in the presence or absence of 1% Triton X-100, then were immunostained using anti-Thy-1 antibody. (A) No primary antibody (control). (B) Immunostained after preincubation for 4 hr in the absence of Triton X-100. (C) A wild-type section was stained with anti-Thy-1 antibody after 4h incubation in the presence of 1% Triton X-100. A Thy1-null section (shown) and 10 wild-type sections (not shown) were preincubated together for 4 hr in 1% Triton X-100, then stained in the presence of 1% Triton X-100. (D) Higher power image of the cerebellum from (C) to emphasize relatively higher staining of the molecular layer (ml), and white matter (wm) compared to the granular layer (gl).

A consequence of these observations is the fact that tissue sections have to be immunostained for gangliosides and GPI-anchored proteins in detergent-free buffers and that all steps must be performed at 4°C. Immunostained sections must be examined and photographed immediately after mounting onto slides because any non-aqua mounting media remaining would allow for transfer of GPI-anchored proteins and glycolipids in between membranes of different cells (unpublished data). Recently, a novel technique called scanning MALDI-quadrupole ion trap time of flight (MALDI-QIT-TOF) imaging mass spectroscopy was used to give a high level of structural information combined with histological accuracy [133].

Taking all mentioned pitfalls of immunochistochemisty together, it seems that a significant amount of previous work on the distribution of these two classes of molecules may need to be re-evaluated.

CONCLUSIONS

The phenomenon of shedding and uptake of gangliosides and GPI-linked proteins have been discovered, forgotten and again discovered several times during the past few decades. In this chapter we have presented evidence from several nearly completely separated scientific areas that clearly demonstrate the ability of GPI-linked proteins to be actively released from the membrane of one cell and inserted in a functional form into membranes of other cells. This phenomenon is difficult to discover and track and it takes a great effort to analyze. Some reasons for this could be various numbers of GPI-linked proteins that have different mechanisms of shedding or that mechanisms could be regulated in a different manner. Shedding could also be very limited in quantity but, on the other hand, it could be directed to a limited number of specific cells and they may be difficult to find in organisms. The process of shedding and uptake of GPI-anchored proteins appears to be regulated, and most probably involves catalytic activity of some proteins that still have to be identified. Functional significance of this phenomenon is not known and it will be very interesting to learn how this complicated process aids in the integration of individual cells into complex organisms.

REFERENCES

[1]   Diep, D.B. et al. Glycosylphosphatidylinositol anchors of membrane glycoproteins are binding determinants for the channel-forming toxin aerolysin. J Biol Chem 273, 2355-2360 (1998).

[2]   Kawagoe, K. et al. Glycosylphosphatidylinositol-anchor deficient mice: implications for clonal dominance of mutant cells in paroxysmal nocturnal hemoglobinuria. Blood 87, 3600–06 (1996).

[3]   Rosti, V. et al. Murine embryonic stem cells without Pig-a gene activity are competent for hematopoiesis with the PNH phenotype but not for clonal expansion. J Clin Invest 100, 1028–1036 (1997).

[4]   Nozaki, M. et al. Developmental abnormalities of glycosylphosphatidylinositol-anchordeficient embryos revealed by Cre/loxP system. Lab Invest 79, 293–299 (1999).

[5]   Abdul-Aziz, N.M. et al. EphrinA-EphA receptor interactions in mouse spinal neurulation: implications for neural fold fusion. Int J Dev Biol 53, 559-568 (2009).

[6]   Ikezawa, H. Glycosylphosphatidylinositol (GPI)-anchored proteins. Biol Pharm Bull 25, 409–417 (2002).

[7]   Eisenhaber, B. et al. Enzymes and auxiliary factors for GPI lipid anchor biosynthesis and post-translational transfer to proteins. BioEssays 25, 367–385 (2003).

[8]   Mayor, S. & Riezman, H. Sorting GPI-anchored proteins. Nat Rev Mol Cell Biol 5, 110–120 (2004).

[9]   Sharom, F.J. & Radeva, G. GPI-anchored protein cleavage in the regulation of transmembrane signals. Subcell Biochem 37, 285–315 (2004).

[10]   Dykstra, M. et al. Location is everything: lipid rafts and immune cell signaling. Annu Rev Immunol 21, 457–481 (2003).

[11]   Allende, D. et al. Jumping to rafts: gatekeeper role of bilayer elasticity. Trends Biochem Sci 29, 325–330 (2004).

[12]   Pike, L. Lipid rafts: heterogeneity on the high seas. Biochem J 378, 281–292 (2004).

[13]   Rao, R. et al. Lipid rafts in cytokine signaling. Cytokine Growth Factor Rev 15, 103–110 (2004).

[14]   Simons, K. & Vaz, W.L. Model systems, lipid rafts, and cell membranes. Annu Rev Biophys Biomol Struct 33, 269–295 (2004).

[15]   Ladisch, S. et al. Shedding and immunoregulatory activity of YAC-1 lymphoma cell gangliosides. Cancer Res 43, 3808–3813 (1983).

[16]   Slein, M.W. & Logan, Jr G.F. Partial purification and properties of two phospholipases of Bacillus cereus. J Bacteriol 85, 369–381 (1963).

[17]   Ikezawa, H. et al. Studies on phosphatidylinositol phosphodiesterase (phospholipase C type) of Bacillus cereus: I. purification, properties and phosphatase-releasing activity. Biochem Biophys Acta 450, 154–164 (1976).

[18]   Futerman, A.H. et al. Identification of covalently bound inositol in the hydrophobic membraneanchoring domain of Torpedo acetylcholinesterase. Biochem Biophys Res Commmun 129, 312–317 (1985).

[19]   Roberts, W.L. & Rosenberry, T.L. Identification of covalently attached fatty acids in the hydrophobic membrane-binding domain of human erythrocyte acetylcholinesterase. Biochem Biophys Res Commmun 133, 621–627 (1985).

[20]   Tse, A.G. et al. A glycophospholipid tail at the carboxyl terminus of the Thy-1 glycoprotein of neurons and thymocytes. Science 230, 1003–1008 (1985).

[21]   Ferguson, M.A. et al. Trypanosoma brucei variant surface glycoprotein has a sn-1,2-dimyristyl glycerol membrane anchor at its COOH terminus. J Biol Chem 260, 4963–4968 (1985).

[22]   Ferguson, M.A. et al. Glycosyl-sn-1,2-dimyristylphosphatidylinositol is covalently linked to Trypanosoma brucei variant surface glycoprotein. J Biol Chem 260, 14547–14555 (1985).

[23] Eisenhaber, B. et al. Post-translational GPI lipid anchor modification of proteins in kingdoms of life: analysis of protein sequence data from complete genomes. Protein Eng 14, 17–25 (2001).

[24] Ferguson, M.A. & Williams, A.F. Cell-surface anchoring of proteins via glycosyl-phosphatidylinositol structures. Annu Rev Biochem 57, 285–320 (1988).

[25] Cross, G.A. Glycolipid anchoring of plasma membrane proteins. Annu Rev Cell Biol 6, 1–39 (1990).

[26] Kinoshita, T. & Inoue, N. Dissecting and manipulating the pathway for glycosylphos-phatidylinositol-anchor biosynthesis. Curr Opin Struct Biol 4, 632–638 (2000).

[27] Ait-Slimane, T. et al. Basolateral internalization of GPI-Anchored proteins occurs via a clathrin-independent flotillin-dependent pathway in polarized hepatic cells. Mol Biol Cell in press (2009).

[28] Miyata, T. et al. The cloning of PIG-A, a component in the early step of GPI-anchor biosynthesis. Science 259, 1318–1320 (1993).

[29] Takeda, J. et al. Deficiency of the GPI anchor caused by a somatic mutation of the PIG-A gene in paroxysmal nocturnal hemoglobinuria. Cell 73, 703–711 (1993).

[30] Bessler, M. et al. Paroxysmal nocturnal haemoglobinuria (PNH) is caused by somatic mutations in the PIG-A gene. EMBO J 13, 110-117 (1994).

[31] Kinoshita, T. et al. Defective glycosyl phosphatidylinositol anchor synthesis and paroxysmal nocturnal hemoglobinuria. Adv Immunol 60, 57–103 (1995).

[32] Rosse, W.F. & Ware, R.E. The molecular basis of paroxysmal nocturnal hemoglobinuria. Blood 86, 3277–3286(1995).

[33] Mortazavi, Y. et al. The spectrum of PIG-A gene mutations in aplastic anemia/ paroxysmal nocturnal hemoglobinuria (AA/PNH): a high incidence of multiple mutations and evidence of a mutational hot spot. Blood 101, 2833–2841 (2003).

[34] Luzzatto, L. et al. Somatic mutations in paroxysmal nocturnal hemoglobinuria: a blessing in disguise? Cell 88, 1–4 (1997).

[35] Barcellini, W. et al. Increased resistance of PIG-A-bone marrow progenitors to tumor necrosis factor a and interferon gamma: possible implications for the in vivo dominance of paroxysmal nocturnal hemoglobinuria clones. Haematologica 89, 651–656 (2004).

[36] Brugger, B. et al. The membrane domains occupied by glycosylphosphatidylinositol-anchored prion protein and Thy-1 differ in lipid composition. J Biol Chem 279, 7530–7536 (2004).

[37] Munro, S. Lipid rafts: elusive or illusive? Cell 2003; 115: 377–388.

[38] Simons K, Toomre D. Lipid rafts and signal transduction. Nat ReV Mo. Cell Biol 1, 31–39 (2000).

[39] Szpurka, H. et al. Altered lipid raft composition and defective cell death signal transduction in glycosylphosphatidylinositol anchor-deficient PIG-A mutant cells. British J of Haematology 142, 413-422 (2008).

[40] Lai, E.C. Lipid rafts make for slippery platforms. J Cell Biol 162, 365–370 (2003).

[41] Varma, R. & Mayor, S. GPI-anchored proteins are organized in submicron domains at the cell surface. Nature 394, 798–801 (1998).

[42] Friedrichson, T. & Kurzchalia, T.V. Microdomains of GPI-anchored proteins in living cells revealed by crosslinking. Nature 394, 802–805 (1998).

[43] Harder, T. et al. Lipid domain structure of the plasma membrane revealed by patching of membrane components. J Cell Biol 141, 929–942 (1998).

[44] Pralle, A. et al. Sphingolipidcholesterol rafts diffuse as small entities in the plasma membrane of mammalian cells. J Cell Biol 148, 997–1008 (2000).

[45] Sharma, P. et al. Nanoscale organization of multiple GPI-anchored proteins in living cell membranes. Cell 116, 577–589 (2004).

[46] Ayuyan, A.G. & Cohen, F.S. Raft composition at physiological temperature and pH in the absence of detergents. Biophysical J 94, 2654–2666 (2007).

[47] Pinaud, P. et al. Dynamic Partitioning of a Glycosyl-Phosphatidylinositol-Anchored Protein in Glycosphingolipid-Rich Microdomains Imaged by Single-Quantum Dot. Tracking Traffic 10, 691–712 (2009).

[48] Wang, T.Y. et al. Artificially lipid-anchored proteins can elicit clustering-induced intracellular signaling events in Jurkat Tlymphocytes independent of lipid raft association. J Biol Chem 280, 22839–22846 (2005).

[49] Douglass, A.D. & Vale, R.D. Single-molecule microscopy reveals plasma membrane microdomains created by protein–protein networks that exclude or trap signaling molecules in T cells. Cell 121, 937–950 (2005).

[50] Low, M.G. Glycosyl-phosphatidylinositol: a versatile anchor for cell surface proteins, FASEB J 3, 1600–1608 (1989).

[51] Braun-Breton, C. et al. Induction of the proteolytic activity of a membrane protein in Plasmodium falciparum by phosphatidyl inositol-specific phospholipase C. Nature 332, 457–459 (1988).

[52] Gmachl, M. et al. The human sperm protein PH- 20 has hyaluronidase activity. FEBS Lett 336, 545–548 (1993).

[53] Brewis, I.A. et al. Activation of the glycosylphosphatidylinositol- anchored membrane dipeptidase upon release from pig kidney membranes by phospholipase C. Biochem J 303, 633–638 (1994).

[54] Lehto, M.T. & Sharom, F.J. Release of the glycosylphosphatidylinositolanchored enzyme ecto-5'-nucleotidase by phospholipase C: catalytic activation and modulation by the lipid bilayer. Biochem J 332, 101–109 (1998).

[55] Tozeren, A. et al. Micromanipulation of adhesion of a Jurkat cell to a planar bilayer membrane containing lymphocyte function-associated antigen 3 molecules. J Cell Biol 116, 997–1006 (1992).

[56] Muller, G. & Bandlow, W. Lipolytic membrane release of two phosphatidylinositol- anchored cAMP receptor proteins in yeast alters their ligandbinding parameters. Arch Biochem Biophys 308, 504–514 (1994).

[57] Wang, X. et al. Variant GPI structure in relation to membrane-associated functions of a murine folate receptor. Biochemistry 35, 16305–16312 (1996).

[58] Kondoh, G. et al. Angiotensin-converting enzyme is a GPI-anchored protein releasing factor crucial for fertilization. Nat Med 11, 160–166 (2005).

[59] Chan, B.L. et al. Insulinstimulated release of lipoprotein lipase by metabolism of its phosphatidylinositol anchor. Science 241, 1670–1672 (1988).

[60] Romero, G. et al. Phosphatidylinositol-glycan anchors of membrane proteins: potential precursors of insulin mediators. Science 240, 509–511 (1988).

[61] Movahedi, S. & Hooper, N.M. Insulin stimulates the release of the glycosyl phosphatidylinositol-anchored membrane dipeptidase from 3T3-L1 adipocytes through the action of a phospholipase C. Biochem J 326, 531–537 (1997).

[62] Park, S.W. et al. Endogenous glycosylphosphatidylinositol-specific phospholipase C releases renal dipeptidase from kidney proximal tubules *in vitro*. Biochem J 353, 339–344 (2001).

[63] Park, S.W. et al. Glycosyl phosphatidylinositol (GPI)-anchored renal dipeptidase is released by a phospholipase C in vivo. Kidney Blood Press Res 25, 7–12 2002).

[64] Davitz, M.A. et al. A glycan-phosphatidylinositol-specific phospholipase D in human serum. Science 238, 81–84 (1987).

[65] Low, M.G. & Prasad, A.R. A phospholipase D specific for the phosphatidylinositol anchor of cell-surface proteins is abundant in plasma. Proc Natl Acad Sci USA 85, 980–984 (1988).

[66] Flores-Borja, F. et al. Genetic regulation of mouse glycosylphosphatidylinositol-phospholipase D. Biochimie 86, 275–282 (2004).

[67] Stieger, S. et al. Enzymatic properties of phosphatidylinositol-glycan-specific phospholipase C from rat liver and phosphatidylinositol-glycan-specific phospholipase D from rat serum. Eur J Biochem 197, 67–73 (1991).

[68] Hoener, M.C. & Brodbeck, U. Phosphatidylinositol-glycan-specific phospholipase D is an amphiphilic glycoprotein that in serum is associated with high-density lipoproteins. Eur J Biochem 206, 747–757 (1992).

[69] Low, M.G. & Huang, K.S. Factors affecting the ability of glycosylphosphatidylinositol- specific phospholipase D to degrade the membrane anchors of cell surface proteins. Biochem J 279, 483–493 (1991).

[70] Mann, K.J. et al. Effect of glycosylphosphatidylinositol (GPI)-phospholipase D overexpression on GPI metabolism. Biochem J 378, 641–648 (2004).

[71] Lehto, M.T. & Sharom, F.J. PI-specific phospholipase C cleavage of a reconstituted GPI-anchored protein: modulation by the lipid bilayer. Biochemistry 41, 1398–1408 (2002).

[72] Moon, Y.G. et al. Conversion of glycosylphosphatidylinositol (GPI)-anchored alkaline phosphatase by GPI-PLD. Arch Pharm Res 22, 249–254 (1999).

[73] Mukasa, R. et al. Characterization of glycosylphosphatidylinositol (GPI)-anchored NCAM on mouse skeletal muscle cell line C2C12: the structure of the GPI glycan and release during myogenesis. Arch Biochem Biophys 318, 182–190 (1995).

[74] Wilhelm, O.G. et al. Cellular glycosylphosphatidylinositol-specific phospholipase D regulates urokinase receptor shedding and cell surface expression. J Cell Physiol 180, 225–235 (1999).

[75] Naghibalhossaini, F. & Ebadi, P. Evidence for CEA release from human colon cancer cells by an endogenous GPI-PLD enzyme. Cancer Lett 234(2), 158-67 (2006).

[76]   Bianco, C. et al. Cripto-1: an oncofetal gene with many faces. Curr Top Dev Biol 67, 85-133 (2005).

[77]   Watanabe, K. et al. Growth factor induction of Cripto-1 shedding by glycosylphosphatidylinositol-phospholipase D and enhancement of endothelial cell migration. J Biol Chem 282, 31643-55 (2007).

[78]   Turner, A.J. & Hooper, N.M. The angiotensin-converting enzyme gene family: genomics and pharmacology. Trends Pharmacol Sci 23, 177–183 (2002).

[79]   Corvol, P. et al. Peptidyl-dipeptidase A/angiotensin Iconverting enzyme. In: Barrett AJ, Rawlings ND, Woessner JF, Eds. Handbook of Proteolytic Enzymes. 2nd Edition, Elsevier, London pp. 332–346 (2004).

[80]   Sun, X. et al. Interaction of angiotensin-converting enzyme (ACE) with membrane-bound carboxypeptidase M (CPM) – a new function of ACE. Biol Chem 389, 1477–1485 (2008).

[81]   Taylor, D.R. et al. Role of ADAMs in the ectodomain shedding and conformational conversion of the Prion protein. J Biol Chem in press (2009).

[82]   Rooney, I.A. et al. GPI-anchored complement regulatory proteins in seminal plasma. An analysis of their physical condition and the mechanisms of their binding to exogenous cells. J Clin Invest 97, 1675–1686 (1996).

[83]   Pan, B.T. & Johnstone, R.M. Fate of the transferrin receptor during maturation of sheep reticulocytes *in vitro*: selective externalization of the receptor. Cell 33, 967–978 (1983).

[84]   Harding, C. et al. Receptor-mediated endocytosis of transferrin and recycling of the transferrin receptor in rat reticulocytes. J Cell Biol 97, 329–339 (1983).

[85]   Raposo, G. et al. B lymphocytes secrete antigen-presenting vesicles. J Exp Med 183, 1161–1172 (1996).

[86]   Brasoveanu, L.I. et al. Melanoma cells constitutively release an anchor-positive soluble form of protectin (sCD59) that retains functional activities in homologous complement-mediated cytotoxicity. J Clin Invest 100, 1248–1255 (1997).

[87]   Zitvogel, L. et al. Eradication of established murine tumors using a novel cell-free vaccine: dendritic cellderived exosomes. Nat Med 4, 594–600 (1998).

[88]   van Niel, G. et al. Intestinal epithelial cells secrete exosome-like vesicles. Gastroenterology 121, 337–349 (2001).

[89]   Wolfers, J. et al. Tumor-derived exosomes are a source of shared tumor rejection antigens for CTL cross-priming. Nat Med 7, 297–303 (2001).

[90]   Blanchard, N. et al. TCR activation of human T cells induces the production of exosomes bearing the TCR/CD3/zeta complex. J Immunol 168, 3235–3241 (2002).

[91]   Vidal, M. et al. Aggregation reroutes molecules from a recycling to a vesicle-mediated secretion pathway during reticulocyte maturation. J Cell Sci 110 (Pt 16), 1867–1877 (1997).

[92]   Fevrier, B. et al. Cells release prions in association with exosomes. Proc Natl Acad Sci USA 101, 9683–9688 (2004).

[93]   Johnstone, R.M. et al. Vesicle formation during reticulocyte maturation. Association of plasma membrane activities with released vesicles (exosomes). J Biol Chem 262, 9412–9420 (1987).

[94]   Ronquist, G. & Brody, I. The prostasome: its secretion and function in man. Biochem Biophys Acta 822, 203–218 (1985).

[95]   Rooney, I.A. et al. Physiologic relevance of the membrane attack complex inhibitory protein CD59 in human seminal plasma: CD59 is present on extracellular organelles (prostasomes), binds cell membranes, and inhibits complement-mediated lysis. J Exp Med 177, 1409–1420 (1993).

[96]   Bouma, S.R. et al. Selective extraction of membrane-bound proteins by phospholipid vesicles. J Biol Chem 252, 6759–6763 (1977).

[97]   Cook, S.L. et al. Cell to vesicle transfer of intrinsic membrane proteins: effect of membrane fluidity. Biochemistry 19, 4601–4607 (1980).

[98]   Medof, M.E. et al. Cellsurface engineering with GPI-anchored proteins. FASEB J 10, 574– 586 (1996).

[99]   Chesebro, B. et al. Anchorless prion protein results in infectious amyloid disease without clinical scrapie. Science 308, 1435–1439 (2005).

[100] McNally, K.L. et al. Cells expressing anchorless prion proteins are resistant to scrapie infection. J Virol 83, 4469–4475 (2009).

[101] Kooyman, D.L. et al. In vivo transfer of GPI-linked complement restriction factors from erythrocytes to the endothelium. Science 269, 89–92 (1995).

[102] Dunn, D.E. et al. A knock-out model of paroxysmal nocturnal hemoglobinuria: Pig-a(−) hematopoiesis is reconstituted following intercellular transfer of GPI-anchored proteins. Proc Natl Acad Sci USA 93, 7938–7943 (1996).

[103] Rifkin, M.R. & Landsberger, F.R. Trypanosome variant surface glycoprotein transfer to target membranes: a model for the pathogenesis of trypanosomiasis. Proc Natl Acad Sci USA 87, 801–805 (1990).

[104] Vakeva, A. et al. High-density lipoproteins can act as carriers of glycophosphoinositol lipid-anchored CD59 in human plasma. Immunology 82, 28–33 (1994).

[105] Suzuki, K. & Okumura Y. GPI-linked proteins do not transfer spontaneously from erythrocytes to liposomes. New aspects of reorganization of the cell membrane. Biochemistry 39, 9477–9485 (2000).

[106] Long, K.E. et al. Timedependent loss of surface complement regulatory activity during storage of donor blood. Transfusion 33, 294–300 (1993).

[107] Butikofer, P. et al. Enrichment of two glycosyl-phosphatidylinositol-anchored proteins, acetylcholinesterase and decay accelerating factor, in vesicles released from human red blood cells. Blood 74, 1481–1485 (1989).

[108] Sloand, E.M. et al. Correction of the PNH defect by GPI-anchored protein transfer. Blood 92, 4439–4445 (1998).

[109] Sloand, E.M. et al. Transfer of glycosylphosphatidylinositol-anchored proteins to deficient cells after erythrocyte transfusion in paroxysmal nocturnal hemoglobinuria. Blood 104, 3782–3788 (2004).

[110] Medof, M.E. et al. Amelioration of lytic abnormalities of paroxysmal nocturnal hemoglobinuria with decayaccelerating factor. Proc Natl Acad Sci USA 82, 2980–2984 (1985).

[111] Babiker, A.A. et al. Transfer of prostasomal CD59 to CD59-deficient red blood cells results in protection against complement-mediated hemolysis. Am J Reprod Immunol 47, 183–192 (2002).

[112] Civenni, G. et al. *In vitro* incorporation of GPI-anchored proteins into human erythrocytes and their fate in the membrane. Blood 91, 1784–1792 (1998).

[113] Walter, E.I. et al. Effect of glycoinositolphospholipid anchor lipid groups on functional properties of decay-accelerating factor protein in cells. J Biol Chem 267, 1245–1252 (1992).

[114] Zhang, F. et al. Spontaneous incorporation of the glycosyl-phosphatidylinositol-linked protein Thy-1 into cell membranes. Proc Natl Acad Sci USA 89, 5231–5235 (1992).

[115] Bulow, R. et al. Rapid lateral diffusion of the variant surface glycoprotein in the coat of Trypanosoma brucei. Biochemistry 27, 2384–2388 (1988).

[116] Morgan, B.P. et al. Cross-linking of CD59 and of other glycosyl phosphatidylinositolanchored molecules on neutrophils triggers cell activation via tyrosine kinase. Eur J Immunol 23, 2841–2850 (1993).

[117] van den Berg, C.W. et al. Exogenous glycosyl phosphatidylinositol-anchored CD59 associates with kinases in membrane clusters on U937 cells and becomes Ca(2+)- signaling competent. J Cell Biol 131, 669–677 (1995).

[118] Premkumar, D.R. et al. Properties of exogenously added GPI-anchored proteins following their incorporation into cells. J Cell Biochem 82, 234 245 (2001).

[119] Metzner, C. et al. Asociation of glycosylphosphatidylinositol-anchored protein with retroviral particles. FASEB J 22, 2734–2739 (2008).

[120] Anderson, S.M. et al. Intercellular transfer of a glycosylphosphatidylinositol (GPI)-linked protein: release and uptake of CD4-GPI from recombinant adeno-associated virus-transduced HeLa cells. Proc Natl Acad Sci USA 93, 5894–5898 (1996).

[121] Keller, G.A. et al. Endocytosis of glycophospholipidanchored and transmembrane forms of CD4 by different endocytic pathways. EMBO J 11, 863–874 (1992).

[122] Lu, P. & Sharom, F.J. Immunosuppression by YAC-1 lymphoma: role of shed gangliosides. Cell Immunol 173, 22–32 (1996).

[123] Deng, W. et al. Influence of cellular ganglioside depletion on tumor formation. J Natl Cancer Inst 92, 912–917 (2000).

[124] Gould, S.J. et al. The Trojan exosome hypothesis. Proc Natl Acad Sci USA 100, 10592–10597 (2003).

[125] Lauc, G. & Heffer-Lauc, M. Shedding and uptake of gangliosides and glycosylphosphatidylinositol-anchored proteins. Biochem Biophys Acta 1760, 584–602 (2006).

[126] Simons, M. et al. Exogenous administration of gangliosides displaces GPIanchored proteins from lipid microdomains in living cells. Mol Biol Cell 10, 3187–3196 (1999).

[127] Crespo, P.M. et al. Effect of gangliosides on the distribution of a glycosylphosphatidylinositol-anchored protein in plasma membrane from Chinese hamster ovary-K1 cells. J Biol Chem 277, 44731–44739 (2002).

[128] Werner, M. et al. Effect of formalin tissue fixation and processing on immunohistochemistry. Am J Surg Pathol 24, 1016–1019 (2000).

[129] Levey, A.I. Use of antibodies to visualize neurotransmitter receptor subtypes. In: Sharif NA, Eds. Molecular Imaging in Neuroscience: A Practical Approach. IRL, Oxford pp. 139– 156 (1993).

[130] Heffer-Lauc, M. et al. Membrane redistribution of gangliosides and glycosylphosphatidylinositol- anchored proteins in brain tissue sections under conditions of lipid raft isolation. Biochem Biophys Acta 1686, 200–208 (2005).

[131] Heffer-Lauc, M. et al. Effects of detergents on the redistribution of gangliosides and GPI-anchored proteins in brain tissue sections. J Histochem Cytochem 55, 805–812 (2007).

[132] Schwarz, A. & Futerman, A.H. Determination of the localization of gangliosides using anti-ganglioside antibodies: comparison of fixation methods. J Histochem Cytochem 45, 611–618 (1997).

[133] Sugiura, Y. et al. Imaging mass spectrometry technology and application on ganglioside study; visualization of age-dependent accumulation of C20-ganglioside molecular species in the mouse hippocampus. PLoS ONE 3, e3232 (2008).

[134] Cheng, P.C. et al. A role for lipid rafts in B cell antigen receptor signaling and antigen targeting. J Exp Med 190, 1549–1560 (1999).

[135] Couet, J. et al. Interaction of a receptor tyrosine kinase, EGF-R, with caveolins. Caveolin binding negatively regulates tyrosine and serine/threonine kinase activities. J Biol Chem 272, 30429–30438 (1997).

[136] Garcia-Cardena, G. et al. Endothelial nitric oxide synthase is regulated by tyrosine phosphorylation and interacts with caveolin-1. J Biol Chem 271, 27237–27240 (1996).

[137] Baird, B. et al. How does the plasma membrane participate in cellular signaling by receptors for immunoglobulin E? Biophys Chem 82, 109–119 (1999).

[138] Mastick, C.C. et al. Insulin stimulates the tyrosine phosphorylation of caveolin. J Cell Biol 129, 1523–1531 (1995).

[139] Wary, K.K. et al. A requirement for caveolin-1 and associated kinase Fyn in integrin signaling and anchorage dependent cell growth. Cell 94, 625–634 (1998).

[140] Janes, P.W. et al. The role of lipid rafts in T cell antigen receptor (TCR) signaling. Semin Immunol 12, 23–34 (2000).

CHAPTER 3

GPI-Anchored Proteins: Biophysical Behaviour and Cleavage by Pi-Specific Phospholipases

Frances J. Sharom

Abstract: Glycosylphosphatidylinositol (GPI) anchors appear to confer unique biophysical properties on the proteins to which they are covalently linked. Model membrane systems provide a powerful tool to explore the effects of bilayer properties on the behaviour of GPI-anchored proteins. Such studies have typically been carried out after reconstitution/insertion of purified GPI-anchored proteins into symmetric or asymmetric lipid bilayer vesicles, supported lipid bilayers, or lipid monolayers. Biophysical studies using atomic force microscopy and Langmuir isotherms have revealed quantitative details of the interactions between GPI-anchored proteins and model membranes. Proteins carrying GPI anchors are believed to be targeted to detergent-resistant cholesterol/ sphingolipid-rich lipid rafts in both intact cells and model membranes, and the special properties of these microenvironments may also modulate their functional activity. GPI-anchored proteins are likely closely associated with the bilayer surface, so that the biophysical properties of the membrane, including curvature and lipid fluidity, modulate their conformation and activity. The GPI anchor can be cleaved by both endogenous and exogenous phosphatidylinositol (PI)-specific phospholipases C and D, from sources such as bacteria, protozoa and mammalian tissues. The release of GPI-anchored proteins in soluble form by phospholipases may play a key role in regulating their surface expression and activity. The GPI anchor appears to impose structural restraints, and its removal may alter the conformation, antigenicity and enzymatic activity of the protein. PI-specific phospholipases must interact closely with the membrane surface to cleave GPI anchors, and their activity is also greatly influenced by membrane biophysical properties.

RECONSTITUTION OF GPI-ANCHORED PROTEINS INTO MODEL MEMBRANE SYSTEMS

Glycosylphosphatidylinositol (GPI)-anchored proteins are unique among membrane proteins in possessing a covalently-linked phospholipid as part of their structure. This lipid moiety is instrumental in anchoring them in the membrane such that their behaviour is difficult to distinguish from that of integral proteins anchored by membrane-spanning polypeptide α-helices. Indeed, many GPI-anchored ectoenzymes were believed to be held in the membrane via transmembrane protein segments until their documented release by phosphatidylinositol (PI)-specific phospholipases forced a re-thinking of this issue. One consequence of the attached GPI anchor is that it confers a unique set of membrane interactions, which may in turn modulate the behaviour and biological activity of the protein, such as its association with lipid raft microdomains. The presence of the GPI anchor, in turn, may influence the properties of the membrane into which the protein is inserted. The best way to explore the biochemical and biophysical basis of these interactions *in vitro* is to reconstitute purified GPI-anchored proteins into lipid bilayer or monolayer model membrane systems with defined physicochemical properties. In reconstituted systems, the experimenter has direct control over the acyl chain length and unsaturation (which control bilayer fluidity), bilayer charge, *etc.*, and other membrane components such as cholesterol and glycosphingolipids can be readily included. The lipid:protein ratio can also be controlled over a wide range, from as low as 2.2:1 (w/w)1 to 200:1 (w/w)2. In the case of lipid bilayers, the choice of reconstitution method and lipid components may also allow some control over the final vesicle size. A wide variety of biophysical techniques may be applied to reconstituted GPI-anchored proteins, including differential scanning calorimetry (DSC), atomic force microscopy (AFM), fluorescence spectroscopy, and Fourier transform infra-red (FTIR) spectroscopy. Such reconstituted bilayers and monolayers are powerful tools to mimic more complex cellular systems and shed light on important biological processes involving proteins that possess GPI anchors in their native state. They may also be useful in the study of proteins that have been genetically manipulated to contain a GPI linkage, and thus possess specific molecular properties.

Symmetric and Asymmetric Lipid Bilayers

The detergent dialysis technique commonly used for transmembrane proteins3 has also been applied to the reconstitution of GPI-anchored proteins and ectoenzymes, such as Thy-1 antigen, lymphocyte function-associated antigen 3 (LFA-3), and 5N-nucleotidase (5N-NTase)(*e.g.* [1,2,4-7]). Typically, the lipid or lipid mixture of choice is completely solubilized in a detergent with a high critical micelle concentration, such as 3- [(3-cholamidopropyl) dimethylammonio]-1-propane-sulfonate (CHAPS) or *n*-octyl-β-D-glucoside (OG). This property of the detergent is essential for its later removal by dialysis. The desired amount of the purified GPI-anchored protein is then added, and after a period of equilibration, the lipid-protein mixture is subjected to dialysis to completely remove the detergent. During this process, unilamellar bilayer vesicles form spontaneously, with the GPI-anchored protein incorporated into them, typically in a roughly symmetrical fashion, with half the protein in the outer leaflet, and the other half in the inner leaflet (Fig. **1C**) [2]. Recovery of GPI-anchored proteins in reconstituted vesicles prepared using this approach is usually >95%. Reconstituted vesicles can also be formed by rapid dilution of small samples of lipid and GPI-anchored protein solubilized in detergent solution (for example, dilution from 40 µL to 1 ml) [8]. Dynamic light scattering (DLS) provides a rapid and convenient method for characterization of the diameter and polydispersity of lipid bilayer vesicles prepared by either method.

The distribution of the GPI-anchored protein between the outer and inner leaflets may be determined by measuring its catalytic activity under permeabilized and non-permeabilized conditions. When the bilayer remains intact and assay reagents cannot penetrate to the lumen, only the enzyme activity of proteins anchored in the outer leaflet is obtained. Addition of permeabilizing concentrations of detergent allows access of the assay reagents to the vesicle interior, so that the activity of the proteins in the inner leaflet is also measured. Typically, an approximate doubling of activity on permeabilization is an indicator that the GPI-anchored protein is symmetrically distributed between the outer and inner membrane leaflets (*e.g.* [2]). Alternatively, for GPI-anchored proteins with no enzymatic activity, cleavage by bacterial PI-specific phospholipase C (PI-PLC; see below) can be used to estimate the symmetry of reconstitution, since >90% of proteins anchored in the outer leaflet are typically released. However, it should be noted that in some GPI-anchored proteins, acylation of the inositol ring leads to resistance to phospholipase cleavage. In a typical large unilamellar vesicle of diameter 100 nm, there are not expected to be any steric hindrance-related issues for proteins anchored in the inner leaflet, since the typical GPI-anchored protein, alkaline phosphatase (AP), is about 5 nm in diameter, and the fully extended anchor is <5 nm in length, so the overall size is <10 nm.

Planar lipid bilayers on solid supports (Fig. **1B**) are very useful for measurements of lateral diffusion and adhesion, and for application of various microscopic imaging techniques [9,10]. Supported lipid bilayers are typically prepared using one of two approaches; by sequential transfer of two lipid monolayers from the air-water interface of a Langmuir-Blodgett trough (see below) to a solid support, such as glass, quartz, mica or a single crystal silicon wafer, or by placing small unilamellar vesicles onto a clean hydrophilic glass surface, where they fuse with each other [11]. Incorporation of fluorescent lipid probes, followed by epifluorescence microscopy, can be used to establish that the films are continuous, and allow rapid lipid diffusion over their surface. In both cases, GPI-anchored proteins can be incorporated into the supported bilayers; in the first case by insertion of protein into the monolayers via injection into the subphase [12] (see below), and in the second by protein reconstitution into the unilamellar vesicles using detergent dialysis [7,13]. The former approach has the advantage of allowing incorporation of the GPI-anchored protein into only the outer monolayer [12].

There has been considerable interest in developing reconstituted vesicle systems with GPI-anchored proteins incorporated into only the outer membrane leaflet (Fig. **1D**), since this would provide an attractive and more faithful mimic of microdomains such as lipid rafts and caveolae. Spontaneous insertion of GPI-anchored AP enzymes from human liver and placenta into liposomes was first reported 20 years ago [14]. When an aqueous solution of the purified proteins (which form micelles or aggregates [15,16]) was added to pre-formed multilamellar liposomes in the absence of detergent, incorporation into the outer leaflet took place over a period of 8 hours. The incorporated enzymes could be completely released by treatment with bacterial PI-PLC, indicating that their GPI moiety was inserted into the membrane. Several later studies made use of this property of spontaneous insertion to study GPI-anchored proteins into liposomal systems (*e.g.* [12,17-19]). Biotechnological approaches can take advantage of the tendency of GPI-anchored proteins to spontaneously insert into membranes to effect the transfer of engineered GPI-anchored molecules into intact cells, a process known as cell "painting" [20].

A detergent-mediated incorporation method previously developed for transmembrane integral proteins [21,22] was used to reconstitute bovine kidney AP into the outer leaflet of liposomes [23]. Purified protein was added to pre-formed phospholipid liposomes in the presence of sufficient detergent (OG was employed) to destabilize the membrane, so that it is positioned just at the onset of the solubilization process [24]. Removal of residual detergent by dialysis led to the formation of liposomes with reconstituted AP located exclusively in the outer membrane leaflet, where it was fully accessible to cleavage by PI-PLC [23]. Interestingly, the presence of AP in the bilayer was observed to alter the stability of the membrane towards detergent solubilization. When a similar detergent-mediated approach was used to reconstitute bovine intestine AP, except that residual detergent was removed using adsorption to hydrophobic resin beads, only 70-85% of the reconstituted protein was in the outer leaflet, with the remaining protein in the inner leaflet facing the vesicle lumen [25]. The yield of incorporation of GPI-anchored proteins was increased at higher temperatures, *e.g.* 38 °C vs. 25 °C [26]. The presence of cholesterol in the pre-formed liposomes also greatly enhanced the incorporation of AP relative to liposomes of phospholipids alone [26]. It seems likely that the observed effect of sterol is related to the affinity of GPI-anchored proteins for cholesterol-rich lipid microdomains (see below). The detergent-mediated reconstitution into dipalmitoylphosphatidylcholine (DPPC) of a mixture of GPI-anchored proteins from *Leishmania* was also reported to be highly dependent on the presence of cholesterol [27]. Thus the process of asymmetric reconstitution using this detergent-mediated approach is quite sensitive to various experimental parameters. Thy-1 was also reconstituted into liposomes using detergent-mediated insertion, and was shown to be sensitive to removal by PI-PLC [28]. High surface densities of GPI-anchored proteins can be achieved using this method; *e.g.* 260 molecules of AP were incorporated per 70 nm vesicle, for a surface coverage of ~65% [29].

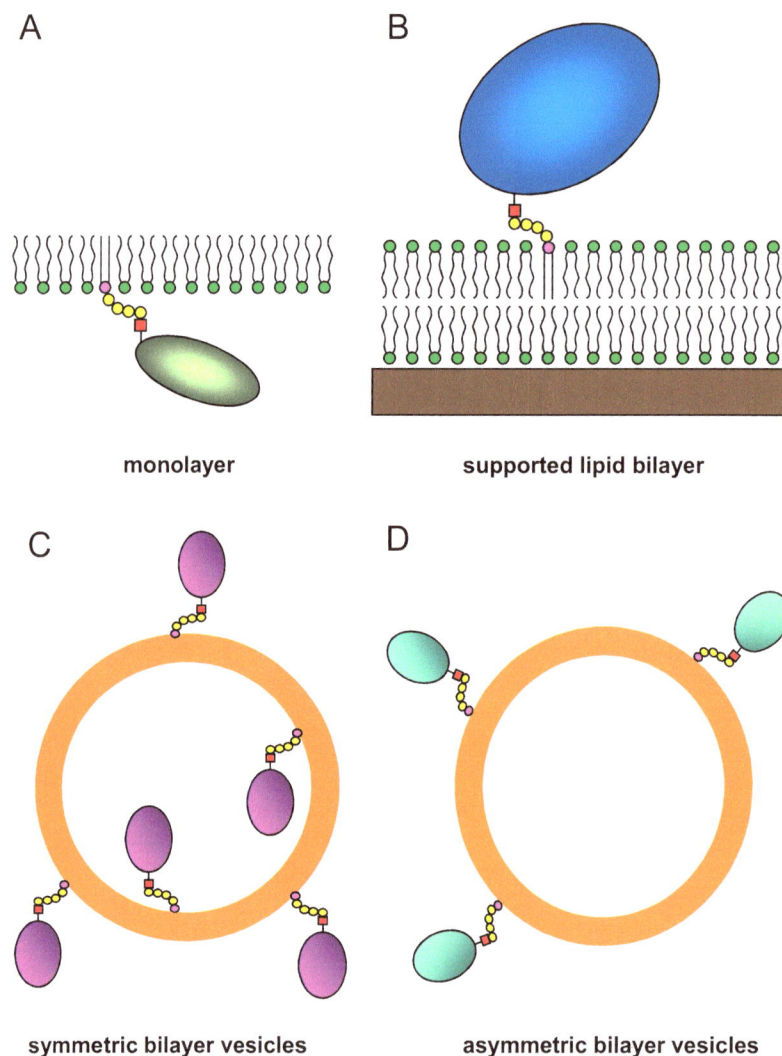

Figure 1: Lipid monolayer and bilayer model systems useful in the study of the biophysical behaviour of GPI-anchored proteins. (A) A phospholipid monolayer spread at the air-water interface, with the lipid headgroups facing the aqueous subphase, and the acyl chains projecting up into the air. GPI-anchored proteins can spontaneously insert into the monolayer when injected into the subphase a few mm below the surface. (B) Lipid bilayers on a solid support (glass, quartz, mica, silicon crystal, *etc.*) containing inserted GPI-anchored proteins. (C) Lipid bilayer vesicles containing symmetrically reconstituted GPI-anchored proteins. The activity of the proteins anchored in the outer leaflet can be detected using membrane-impermeant assay reagents, while detection of the activity of the proteins anchored in the inner leaflet requires detergent permeabilization of the bilayer. Only the proteins

anchored in the outer leaflet can be removed by treatment with PI-PLC. (D) Lipid bilayer vesicles containing GPI-anchored proteins reconstituted asymmetrically into only the outer leaflet. The enzymatic activity of all the proteins can be detected using membrane-impermeant assay reagents, and they can all be cleaved by PI-PLC.

Spontaneous insertion of unconjugated (free) GPI lipids into lipid bilayers was achieved by addition of a small volume of methanol solution containing the lipids (5% final concentration) to pre-formed 100 nm extruded liposomes [30]. Asymmetric localization of GPI lipid in the outer monolayer was shown by its high level of availability for enzymatic hydrolysis by α-galactosidase, which was not increased after detergent permeabilization of the liposomes. The process of GPI lipid insertion led to some leakage of the liposome contents, likely as a result of molecular rearrangements within the bilayer during the insertion process. GPI lipids asymmetrically incorporated into lipid vesicles can also be cleaved by both PI-PLC and GPI-specific phospholipase D (GPI-PLD) [31].

Monolayers

Monolayers at the air-water interface are a useful biomimetic system for studying the interaction of proteins with membranes [32]. Amphiphilic molecules, such as phospholipids, are spread at the air-water interface in a Langmuir trough apparatus, and the resulting monomolecular films can be compressed laterally using sliding barriers. The surface pressure can be measured and controlled, and surface densities and surface packing can be determined. The incorporation into (or surface adsorption onto) the monolayer of various proteins (Fig. **1A**) can provide useful information on their biophysical behaviour in a membrane environment. In addition, the monolayers may be observed using techniques such as Brewster angle microscopy (BAM) and epifluorescence microscopy, to provide information on protein lateral distribution and domain formation.

Early experiments on GPI-anchored proteins used purified bovine intestine AP spread at the buffer/air interface in the absence of lipids, where the protein molecules form a monolayer while maintaining their native state and enzymatic activity [33]. The monolayer could be highly compressed without leading to expulsion of AP into the bulk aqueous phase, indicating that the protein forms a highly stable interfacial film. BAM images showed that prolonged incubation at high surface pressures led to the reversible formation of condensed protein clusters, which remained stable for tens of seconds after decompression. A thermodynamic study of bovine intestine AP showed that the GPI-anchored protein behaved very differently from the same protein lacking the anchor, and presented a model for its interfacial molecular organization [34]. However, a study of rat osseous plate AP found little difference between the two forms in terms of their adsorption to the air-water interface, suggesting that AP enzymes from different tissues may behave differently [35].

The use of Langmuir phospholipid monolayers at the air-water interface provides a means to examine the lipid interactions of GPI-anchored proteins in a host lipid with defined properties, which might mimic their behaviour in the outer leaflet of the plasma membrane. This monolayer system also has the advantage of allowing the experimenter to control the lipid lateral packing density. Injection of bovine intestine AP into the aqueous subphase beneath the monolayer led to an increase in molecular area as the GPI anchor of the protein inserted into the lipids, which became less condensed [36]. A series of studies using rat osseous plate AP in negatively charged monolayers of dimyristoylphosphatidic acid (DMPA) found that the enzymatic activity declined as the surface density of the GPI-anchored protein increased, suggesting that steric effects may hinder the accessibility of the active site to substrate [37].

Caseli and co-workers developed three different methods for incorporating a GPI-anchored protein into DMPA monolayers: (i) injection into the aqueous subphase a few mm below the monolayer surface, (ii) co-spreading, in which the protein solution is spread on the interface immediately after the phospholipid is spread, and (iii) the subphase formation method, in which the protein is already present in the subphase before the phospholipid solution is spread [38]. The surface pressure vs. area curves for monolayers formed using the three different approaches showed that the orientation of the GPI-anchored proteins resulting from the co-spreading and subphase formation methods appeared to be different from that attained using the subphase injection technique. Fluorescence microscopic observations showed that injection led to a much more homogeneous lateral distribution of GPI-anchored protein, whereas the other two methods produced regions that were either devoid of protein or enriched in aggregated protein [38]. This poor surface miscibility effect might result from protein denaturation; the injection technique is thus preferred.

Using mixed monolayers of rat osseous plate AP and DMPA prepared by co-spreading, the maximum enzymatic activity was observed at a surface pressure of 18 mN/m, and decreased greatly at values above this, which correlated with a sharp drop in activity at higher surface densities of the GPI-anchored protein39. Fluorescence microscopy using labelled proteins incorporated into the monolayer showed the existence of protein-rich phases at surface pressures up to 18 mN/m. It was proposed that such protein-rich clusters are responsible for the drop in enzyme activity, perhaps as a result of reduced access of substrate to the active site due to steric hindrance [39]. Thus the lateral organization of GPI-anchored proteins in native membranes may modulate their catalytic activity, especially if they are clustered in microdomains.

The interaction of bovine intestine AP with lipids containing saturated (DPPC) and unsaturated (palmitoyloleoylphosphatidylcholine, POPC) acyl chains showed that both monolayers prevented GPI-protein insertion from the subphase at high initial pressures, due to the high lipid packing density [40]. Both DPPC and DOPC monolayers showed a shift to higher molecular areas after insertion of the protein. BAM images were used to follow the progressive integration of the protein into the monolayer. Evidence was presented suggesting strong interactions between the saturated acyl chains of the GPI anchor with the acyl chains of DPPC, but not POPC. More recently, the incorporation of cholesterol into DPPC monolayers was found to accelerate the insertion and subsequent stabilization of *Leishmania* GPI-anchored proteins, possibly by the formation of raft-like domains [27].

BIOPHYSICAL BEHAVIOUR OF GPI-ANCHORED PROTEINS IN MEMBRANES

Application of AFM and Other Biophysical Techniques

The AFM technique uses a scanning probe microscope, which generates topographical information about a surface by dragging a very sharp probe across it in a raster fashion. An image of the surface is built up from the vertical deflections of the probe as it encounters objects in its path. Rapid advances in AFM technology over the past 10 years mean that these instru-ments can now produce very high resolution images (lateral resolution of 5 Δ and vertical resolution of 1-2 Δ) of soft biological surfaces in a fluid medium, under near-physiological con-ditions. Unique information can also be obtained on the forces of interaction between surface molecules, and even the forces required to unfold proteins. One caveat with the use of AFM is that low force must be used during the scan (<250-300 pN), to avoid extraction of GPI-anchored proteins from the bilayer [41]. In addition, the time required to conduct a surface scan limits the time-resolution for rapidly diffusing proteins at low concentrations. AFM has proved particularly useful for studying the association of purified membrane proteins (including GPI-anchored pro-teins) with supported lipid bilayers [42], and also for imaging synthetic lipid raft microdomains [43].

Substantial work has been carried out using the AFM technique on AP enzymes from different sources (reviewed in [44]). Spontaneously inserted GPI-anchored protein molecules were clearly visualized as dots projecting from the surface of the bilayer. AP clusters were also evident, and pre-treatment of the protein with PI-PLC led to the absence of AP spots in the image, as expected. The preference of GPI-anchored proteins for localization in lipid raft microdomains has also been illustrated at the molecular level using AFM. Because of the increased thickness of rafts domains in the bilayer compared to the fluid regions around them, they are readily visible by AFM as "islands" projecting ~1 nm above the bilayer surface (Fig. **2**). Single molecules of GPI-anchored proteins, such as AP, can readily be visualized protruding from the upper surface of the raft domains [43]. The detailed information provided by AFM has led to important conclusions on the preference of GPI-anchored proteins for distributing between various bilayer domains of differing fluidity and lipid composition. AFM can also explore molecular interactions at the level of force measurements. By covalently tethering AP to an AFM tip, it was possible to quantitate the adhesive interactions of the protein with a supported DPPC bilayer [45]. The AP-functionalized tip was first brought close to the bilayer and the GPI anchor was allowed to penetrate and insert into it. Subsequent separation of the membrane and the tip resulted in removal of the GPI-anchored protein from the membrane (this process was repeated hundreds of time at different locations in the bilayer). The measured force gradually increases until a rupture event takes place, which corresponds to removal of a single AP molecule from the bilayer. The adhesion frequency was increased by a factor of 2 when the tip was functionalized with the GPI-anchored protein, and a rupture force of 200 pN was estimated for removing the four acyl chains of a GPI-anchored AP dimer from the membrane.

The surface of monolayers and supported lipid bilayers can be imaged using epifluorescence microscopy (which entails the use of fluorescently labelled lipids or proteins) and BAM (which does not). Both provide useful information on the organization of GPI-anchored proteins and lipid domains (for example, see [33,38-40]).

Cryoelectron microscopy has also been employed to visualize the lateral distribution of GPI-anchored proteins in a liposomal membrane. For example, reconstituted vesicles containing bovine intestine AP showed a non-homogeneous distribution of the enzyme on their surface, suggesting that the proteins were clustered [29], in agreement with results from AFM.

Fourier transform infrared reflection/absorption spectroscopy (FT-IRRAS) has also been used to study GPI-anchored proteins at membrane surfaces. This technique can identify amide bonds and establish whether the amino acid residues are present in α-helices or β-strands, as well as being able to detect changes in protein secondary structure. In a study of monolayers of pure rat intestine AP at the buffer/air interface, the formation of condensed protein domains was noted at high surface pressures, suggesting the existence of strong attractive interactions between AP molecules in these pure protein films. Such clustering might contribute to concentration of these proteins in microdomains. Spectral changes indicated that the protein was positioned with the GPI anchor standing upright facing the air, and the ellipsoid protein lying parallel to the surface in the aqueous phase [33]. FT-IRRAS of AP after insertion from the aqueous subphase into monolayers of two different phospholipids showed that the protein had a very similar orientation [36]. Comparable results were obtained for AP from rat osseous plate in pure protein monolayers and phospholipid monolayers [46].

Association of GPI-Anchored Proteins with Lipid Raft Microdomains in Model Systems

The plasma membrane displays lateral inhomogeneity, and is believed to contain specialized microdomains with different lipid and protein composition. In 1997, Simons and Ikonen proposed the existence of dynamic assemblies known as lipid rafts [47], which were envisaged as discrete ordered domains enriched in sphingolipids and cholesterol (Fig. **2**). Rafts are capable of segregating specific membrane proteins, thus serving as platforms or "hot-spots" for important cellular processes such as signal transduction [48]. The differential miscibility of the constituent lipids was proposed to be the basis for the formation of rafts, and leads to lateral phase separation into a fluid, liquid crystalline, disordered (ld) phase, and a more tightly packed, liquid ordered (lo) phase rich in cholesterol and sphingolipids. Rafts have been identified with the lo phase, which was first characterized in lipid bilayer model systems. GPI-anchored proteins are believed to associate preferentially with rafts because of the predominantly saturated acyl chains of their anchors (Fig. **2**). It has been difficult to prove the presence of lipid rafts containing GPI-anchored proteins in intact cells; there is evidence both for and against their existence, and the field remains controversial [49,50].

Model systems have been more amenable to both biophysical and biochemical studies of raft association of GPI proteins *in vitro*, especially since the system parameters can be systematically manipulated. Various "raft-containing" lipid mixtures, usually composed of sphingolipids, cholesterol and an unsaturated phosphatidylcholine (PC), have been employed to explore the partitioning of GPI-anchored proteins between liquid ordered ("raft") and fluid ("non-raft") domains under various conditions. Several GPI-anchored proteins, including AP and 5N-NTase, have been shown to associate preferentially with detergent-resistant "raft-like" fractions in reconstituted lipid bilayer vesicle systems (for example, see [5,17,51,52]).

Supported lipid bilayers have also proved to be useful in studying targeting of GPI-anchored proteins to various membrane domains, where they can be visualized by AFM. When bovine intestine AP was added to pre-formed supported bilayers composed of two phase-separated phospholipids, dioleoylphosphatidylcholine (DOPC) and DPPC, the spontaneously inserted GPI-anchored protein molecules were visualized as dots projecting from the surface of the DPPC-enriched domains, which are in the solid gel phase [53]. Thus the presence of the GPI anchor favoured insertion into ordered domains of higher bilayer thickness. A later AFM study found that bovine intestine AP inserted spontaneously into the gel phases of three different PC-sphingomyelin (SM) binary lipid mixtures [41], again suggesting that where two phases of different fluidity co-exist, the GPI-anchored protein strongly prefers the most ordered phase. Insertion of AP was a slow process, taking from 30 min to several hours. In DOPC/SM bilayers, AP molecules were observed at the boundaries of the gel and fluid domains, as well as in the gel phase. AP insertion resulted in net transfer of lipids from the fluid to the gel phase, or the reverse, depending on the lipid mixture. Human placental alkaline phosphatase (PLAP) reconstituted into DOPC/SM (1:1 w/w) supported bilayers with and without cholesterol was also found to partition primarily (~90%) into raft domains, where it was easily visualized by AFM [54]. In contrast, <30% of the fluorescently-tagged PLAP was found in the lo phase in giant unilamellar vesicles with a similar lipid composition [51]. The factors responsible for the different results are not clear, but could be related to the presence of the support, or the different methods used for membrane insertion of the GPI-anchored protein.

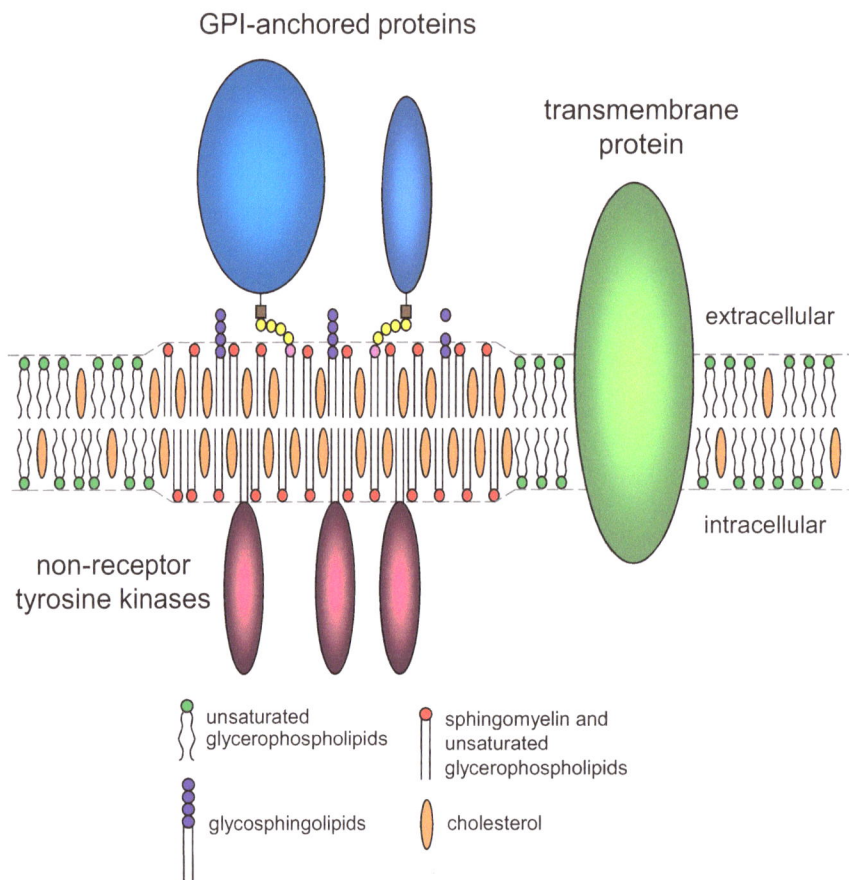

Figure 2: Localization of GPI-anchored proteins in sphingolipid/cholesterol-rich raft microdomains which are thought to be in the l_o phase. Sphingolipids, glycosphingolipids, saturated glycerophospholipids and cholesterol are tightly packed within the rafts, with extended lipid chains, resulting in an increase in membrane thickness that can be observed by AFM. GPI-anchored proteins are inserted into the outer leaflet of the raft domains, while acylated proteins involved in signalling are found in the inner leaflet. Unsaturated glycerophospholipids and many transmembrane proteins are excluded from the raft domains.

The preference of GPI-anchored proteins for associating with raft microdomains in model systems appears to depend on both lipid composition, and the structural characteristics of the particular GPI anchor. Anchors may possess different fatty acyl chain lengths/unsaturation, they may have diacyl or dialkyl constituents, and the inositol ring may also be acylated. All these factors may play a role in the preferred distribution of a particular GPI-anchored protein between fluid and ordered membrane domains, and may also explain why different proteins localize to different subsets of raft microdomains. Factors such as these may explain, for example, the differences in behaviour noted for AP from bovine intestine and human placenta (see above). Temperature may also play a role in GPI-anchored protein distribution between phases. In POPC/SM/cholesterol supported bilayers, AFM showed that bovine intestine AP inserted into the SM/cholesterol ordered domains at lower temperatures, but redistributed into both fluid POPC-enriched and ordered domains above 30°C [55]. It is likely that very few GPI-anchored proteins are located exclusively in raft domains. Most have a preference, such they partition more into the lo domains than the bulk fluid ld domains, or *vice versa*. Even for proteins that greatly prefer to be located within rafts, molecules can still be found in the bulk lipid. However, the local concentration or density of a raft-preferring protein will be much higher within the raft than in the bulk lipid, since the rafts may only occupy a fraction of the membrane surface.

Lateral Mobility and Dynamic Behaviour of GPI-Anchored Proteins

The lateral mobility of cell surface proteins varies over 3 orders of magnitude (lateral diffusion coefficient, D = 10-

11-10-8 cm2/s). Because the membrane-inserted portion of GPI-anchored proteins is a simple phospholipid, it was not surprising that some of them were found to have a high rate of lateral diffusion in intact cells compared to transmembrane proteins (*e.g.* [56]). However, not all proteins with GPI anchors have high lateral diffusion rates; PH20 (a sperm surface antigen) [57] is a notable example. In addition, the extracellular domain of some proteins appears to slow them down, possibly by interacting strongly with other membrane components [58]. Later work examined the lateral mobility of purified GPI-anchored proteins in various model systems. Fluorescence recovery after photobleaching (FRAP) experiments showed that a GPI-anchored version of LFA-3 (an adhesion receptor) reconstituted into supported lipid bilayers had a D value approaching that of a phospholipid, while the transmembrane isoform was immobile (likely because of interactions with the glass surface) [7]. A high rate lateral diffusion of GPI-anchored receptors appeared to enhance adhesion of intact cells to the supported bilayer surface, possibly by allowing rapid accumulation of GPI-linked LFA-3 at contact areas, or by increasing the rate of receptor-ligand interactions on the membrane surface. When GPI-anchored proteins were incorporated into supported lipid bilayers and coupled to 30 nm fluorescent beads, their motion could be tracked by video microscopy [59]. The diffusion coefficients for two GPI-anchored proteins were within a factor of 2 of that for a phospholipid, again showing their intrinsically high mobility. The use of a novel approach to manipulate GPI-anchored proteins in supported lipid bilayers was reported, involving the use of electric fields and diffusion barriers [13]. Three different proteins were used (two genetically engineered, one native, tagged with fluorescent antibodies), together with a fluorescent anionic lipid probe. Under the influence of an electric field, epifluorescence microscopy showed that the GPI-anchored proteins moved towards the cathode and the lipid probe to the anode, in a reversible fashion. Patterns scratched in the membrane acted as barriers to lateral motion, and could be used to "corral" the proteins into regions where they were concentrated 5- to 50-fold. Potential applications of this approach include the forced 2-dimensional crystallization of proteins for structure determination by electron or X-ray crystallography.

Recently it has become possible to produce synthetic GPI anchors of different structure, linked to green fluorescent protein (GFP), and to study their dynamic properties in supported lipid bilayers. Using fluorescence correlation spectroscopy, these proteins were found to diffuse as rapidly as a phospholipid (D = ~3 × 10-8 cm2/s) in supported lipid bilayers [60]. After their insertion into intact cells, FRAP experiments showed that GFP proteins with synthetic anchors were highly mobile, however, they diffused 10-fold more slowly than in supported bilayers, likely because of interactions with other cell surface molecules [61]. Compared to native GPI-anchored proteins, lower D values were found for synthetic anchors in which the core glycan was replaced by a highly flexible linker. This suggests that the more rigid core of a native GPI anchor may reduce non-specific contacts of the protein moiety with other membrane components, and thus enhance mobility.

The lateral motion of GPI-anchored proteins appears to be altered when they are present in lipid raft microdomains. Sheets *et al.* labelled Thy-1 molecules on the surface of intact cells with colloidal gold-conjugated antibodies, and then used single particle tracking to follow them around on the cell surface [62]. 35-40% of the Thy-1 proteins were trapped in transient confinement zones of ~300 nm diameter for 7-9 s intervals, as was a similar fraction of GM1 ganglioside. This confined lateral diffusion likely arises from the presence of both Thy-1 and GM1 in lipid raft microdomains. However, a later study found no appreciable confinement of GPI-anchored MHC Class II molecules relative to the transmembrane isoforms [63]. Quantum dots represent excellent probes for single molecule imaging and tracking in live cells. The lateral movement of GPI-anchored avidin molecules was tracked by labelling with biotin quantum dots, and they were observed to co-localize with GM1 ganglioside in regions of the membrane where they diffused very slowly [64]. There was an instantaneous change in D value for the GPI-anchored protein on entry into, or exit from, these microdomains, which were dependent on cholesterol, but not the actin cytoskeleton. Kusumi and co-workers have pioneered the use of high speed (25 μs), high resolution optical microscopy to follow single molecules around on the membrane surface [65]. They observed that, like membrane phospholipids, both a GPI-anchored protein and its peptide-anchored isoform undergo rapid "hop diffusion" between adjacent 40 nm membrane compartments. These appear to be created by the presence of integral proteins aligned along the membrane skeleton, thus forming a barrier to free lateral diffusion [66]. They found no evidence for transient confinement of GPI-anchored proteins of the type reported by other groups, suggesting that if raft association exists, it is short-lived. The conflicting reports in this field await resolution.

Interaction of GPI-Anchored Proteins with the Membrane Surface

Studies on several different GPI-anchored proteins indicate that the GPI anchor may hold the protein moiety near the membrane surface, thus allowing a close interaction to take place. Homans *et al.* suggested that the glycan moiety of the trypanosomal variant surface glycoprotein (VSG) GPI anchor lay in an extended conformation along the plane of the membrane [67], thus allowing for close interaction of the protein with the surface. Later work on the GPI anchor of Thy-1 suggested that the anchor was in a tightly folded conformation between the lipid surface and the protein [68]. DSC studies on reconstituted Thy-1 antigen showed that each protein molecule perturbed about 53 phospholipids, many more than the 12-18 predicted for the size of a GPI anchor, which suggested that the protein moiety interacts with the membrane surface to perturb additional lipid molecules1. This type of close membrane interaction is not possible if the GPI anchor adopts a fully extended conformation ("lollipop" model, Fig. **3A** for PLAP), which would hold the protein ~33 Δ above the membrane surface [69]. This suggests that the protein portion may flop down onto the membrane with the GPI anchor folded up beneath it, allowing direct contact with the bilayer (Fig. **3B**) [70]. The vertically extended anchor conformation may be induced by steric crowding of GPI-anchored proteins at high protein:lipid ratios, resulting in a change in biophysical behaviour, as observed with reconstituted Thy-11.

More direct evidence for close association of GPI-anchored proteins with the membrane surface comes from fluorescence resonance energy transfer (FRET) studies [8]. Purified PLAP and Thy-1 proteins labelled with a donor fluorophore were reconstituted into bilayers containing a phospholipid labelled in the headgroup with an acceptor fluorophore. Results indicated that the protein moiety for both PLAP and Thy-1 approached the bilayer phospholipid headgroups at a distance of 10-14 Δ, thus possibly resting on the membrane surface.

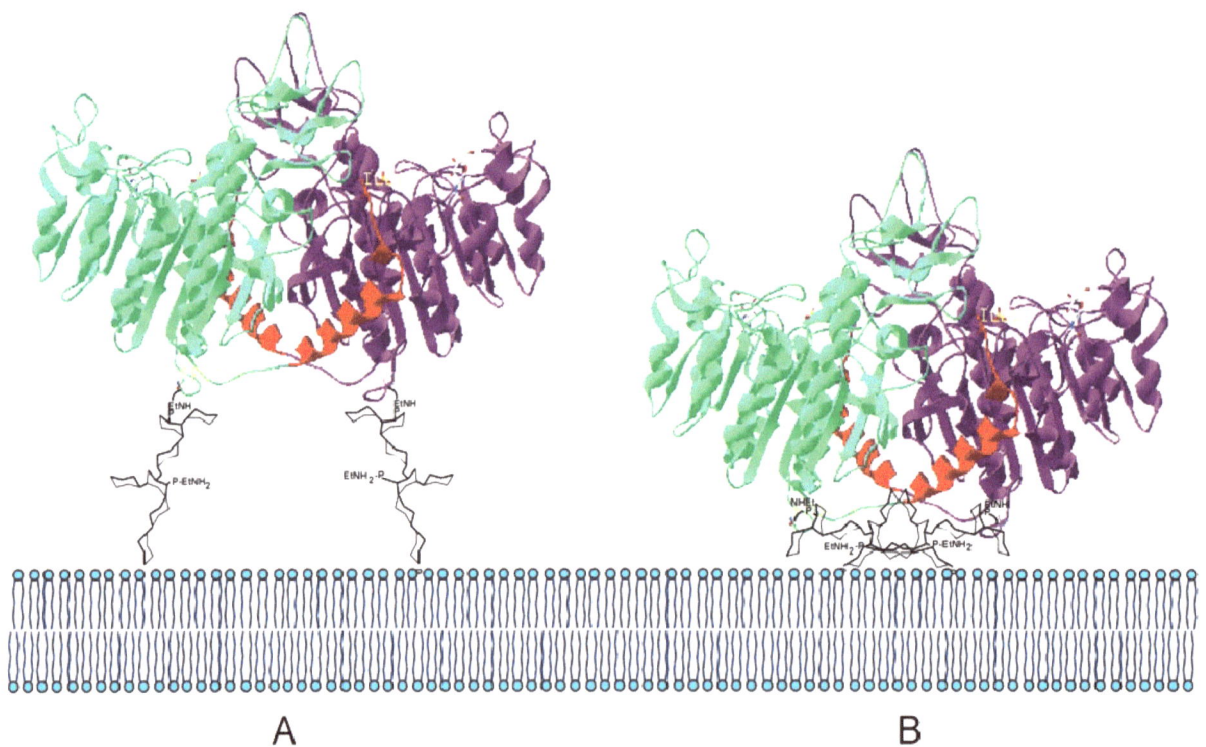

Figure 3: Models for the interaction of dimeric PLAP with the membrane surface. (A) Model of the crystal structure of PLAP with the GPI anchor in a fully extended conformation, holding the protein some distance from the surface of the bilayer. (B) The GPI anchor is shown folded up underneath the protein, allowing PLAP to make contact with the membrane surface. In this conformation, the fluidity and other physicochemical properties of the membrane are able to modulate the enzymatic activity of the protein.

GPI ANCHOR CLEAVAGE BY ENDOGENOUS AND EXOGENOUS PI-SPECIFIC PHOSPHOLIPASES

GPI-anchored proteins can be released from the cell surface in soluble form by specific phospholipase enzymes of the C and D types (PLC and PLD). This cleavage process was instrumental in the initial discovery of the GPI anchor, and remains an important tool for experimental manipulation of this class of proteins. Cleavage of mammalian GPI-anchored proteins by endogenous phospholipases results in down-regulation of their surface expression, and simultaneous up-regulation of extracellular soluble forms in plasma and other body fluids. This process may regulate the functions of these proteins, including their role in signal transduction. Bacterial and protozoan phospholipases have been relatively well characterized, while much less is known about the endogenous enzymes in higher eukaryotes.

Bacterial Phospholipase C

PI-PLC enzymes are secreted extracellularly by several bacterial species, including *Bacillus cereus*, *Bacillus thuringienesis*, *Listeria monocytogenes*, *Clostridium novii* and *Staphylococcus aureus* [71], where they have been linked to pathogenicity [72,73]. These enzymes have been used extensively to cleave mammalian GPI anchors, and were very useful in aiding their structural characterization [74]. The mechanism of cleavage of phosphatidylinositol (PI) and GPI anchors by PI-PLC involves two steps; initial production of diacylglycerol and *myo*-inositol-1,2-cyclic phosphate (cIP), followed by ring opening of the latter to produce *myo*-inositol-1-phosphate (Fig. **4A**). The second hydrolysis step is slow, leading to accumulation of the cyclic intermediate, which forms an epitope known as the cross-reacting determinant (CRD). Specific antibodies can detect the CRD [75], which is used as a diagnostic test for PLC cleavage of a GPI-anchored protein, since PLD operates by a different mechanism, and does not produce cIP. Certain covalent modifications of the inositol ring, especially acylation, can block GPI anchor cleavage by PI-PLC [76,77]. High resolution X-ray structures are known for PI-PLCs from *L. Monocytogenes* [78] (PDB 1A0D) and *B. cereus* [79,80]; structures of the latter exist for the enzyme in an unliganded form (PDB 1PTD), and also complexed to *myo*-inositol (PDB 1PTG) and glucosaminyl(α1→6)-D-*myo*-inositol (PDB 1GYM). The PI-PLC from *B. thuringiensis* has a very high level of sequence identity to the *B. cereus* protein (only 8 of 298 amino acids are different), so their structure and mechanism are likely to be almost identical. In order to cleave their substrates, bacterial PI-PLCs must come into close contact with the membrane and undergo interfacial activation [81]. This is mediated by an interfacial binding surface (IBS) containing a short helix and a loop [82], which are located on the rim of the $\alpha\beta$-barrel structure just outside the active-site cleft (Fig. **4B**). Aromatic and hydrophobic residues within the IBS are thought to be important in surface binding and activation of the phospholipase [83]. As well as cleaving the GPI anchors of proteins, bacterial PI-PLCs can also cleave GPI lipids [31,84], which are present in small amounts in the plasma membrane of cells [85]. The resulting soluble inositolphosphoglycans (IPGs) are believed to be putative second messengers involved in signalling by insulin and other hormones [86].

Protozoan Phospholipases

Several parasitic protozoans, including *Trypanosoma brucei* and *Leishmania* species, express endogenous PI-PLCs. These enzymes transform membrane-bound GPI-anchored proteins such as VSG (*T. brucei*) and gp63 (*Leishmania*) into soluble hydrophilic forms. Trypanosomal PI-PLC operates by the same two-step mechanism as bacterial PI-PLC. The enzyme appears to be important for infection and pathogenicity, although its exact physiological role is unclear, and its loss impairs growth of the parasite in mammalian cells. Trypanosoma's protective surface coat of VSG undergoes cyclical antigenic variation to evade the host immune response, and the PI-PLC may be involved in removal of the coat in the human or tsetse fly hosts.

Mammalian Phospholipases C and D

Soluble forms of several GPI-anchored proteins, including Thy-1, AP, 5N-NTase and carcinoembryonic antigen (CEA), are found extracellularly, suggesting that they are generated by the action of endogenous PLC/PLD enzymes. However, little is known about these enzymes, and their identity and mode of action remain poorly understood. GPI-PLD activity has been found in mammalian serum [87,88] and a number of other tissues [89-92]. The serum enzyme appeared to be released into the circulation following synthesis by the islet cells of the pancreas, and has been purified, cloned and characterized [93]. Unlike bacterial PI-PLC, the GPI-proteins cleaved by GPI-PLD do not cross-react with anti-CRD antibodies, indicating that the enzyme does not proceed through a cIP intermediate. In consequence, GPI-PLD is able to cleave modified acylated GPI anchors that are resistant to PI-PLC cleavage. However, it soon became clear that GPI-PLD is not able to cleave GPI-anchored proteins on native membrane surfaces. It was found to require either the presence of detergent [94], or intracellular activation by proteases [95], to be cleavage-competent. It was later reported that GPI-PLD acquired cleavage activity against native GPI-anchored proteins following post-translational glycosylation [96], suggesting that the immature form undergoes a conformational change after intracellular processing. Of interest, serum GPI-PLD was reported to cleave free GPI lipids from liposomal membranes in the absence of detergent [31,84], suggesting a possible role in signal transduction via IPGs. If free GPI lipids are phosphorylated by PI-3-kinase, this improves their cleavage from the outer surface of liposomes by bacterial PI-PLC, but reduces their cleavage by GPI-PLD [97]. To date, the role of the PLD enzyme in the metabolism of GPI anchors remains unclear.

Proving the existence of endogenous mammalian PI-PLC enzymes has been difficult, especially since they likely co-exist with serum GPI-PLD. The use of anti-CRD antibodies has been helpful in demonstrating that a particular protein was cleaved by a PI-PLC enzyme, rather than by GPI-PLD. Soluble forms of 5N-NTase were shown to arise from cleavage of a GPI-anchored membrane-bound protein using this approach [98]. Renal dipeptidase released spontaneously from kidney proximal tubules and found in urine also arises from a membrane-bound protein by endogenous PI-PLC cleavage [99,100], as does membrane dipeptidase after insulin treatment [101]. However, these PI-PLCs were not identified or characterized. More recently, a genetic screen in *Drosophila* indentified a factor known as Notum, which was able to release GPI-anchored glypicans (a family heparin sulphate proteoglycans) from the cell surface [102]. In a breakthrough for the mammalian PI-PLC field, mouse Notum was expressed, purified and characterized by Filmus and co-workers [103]. The protein indeed proved to be a secreted PI-PLC with cleavage activity against several glypicans and some other GPI-anchored proteins, including urokinase-type plasminogen activator receptor (uPAR) and T-cadherin. Notum was unable to cleave the GPI anchor from CEA, suggesting that it has some sort of specificity, however, its physiological substrates are not known. To date, Notum remains the only characterized endogenous mammalian phospholipase that can cleave GPI anchors on the extracellular surface of native cell membranes.

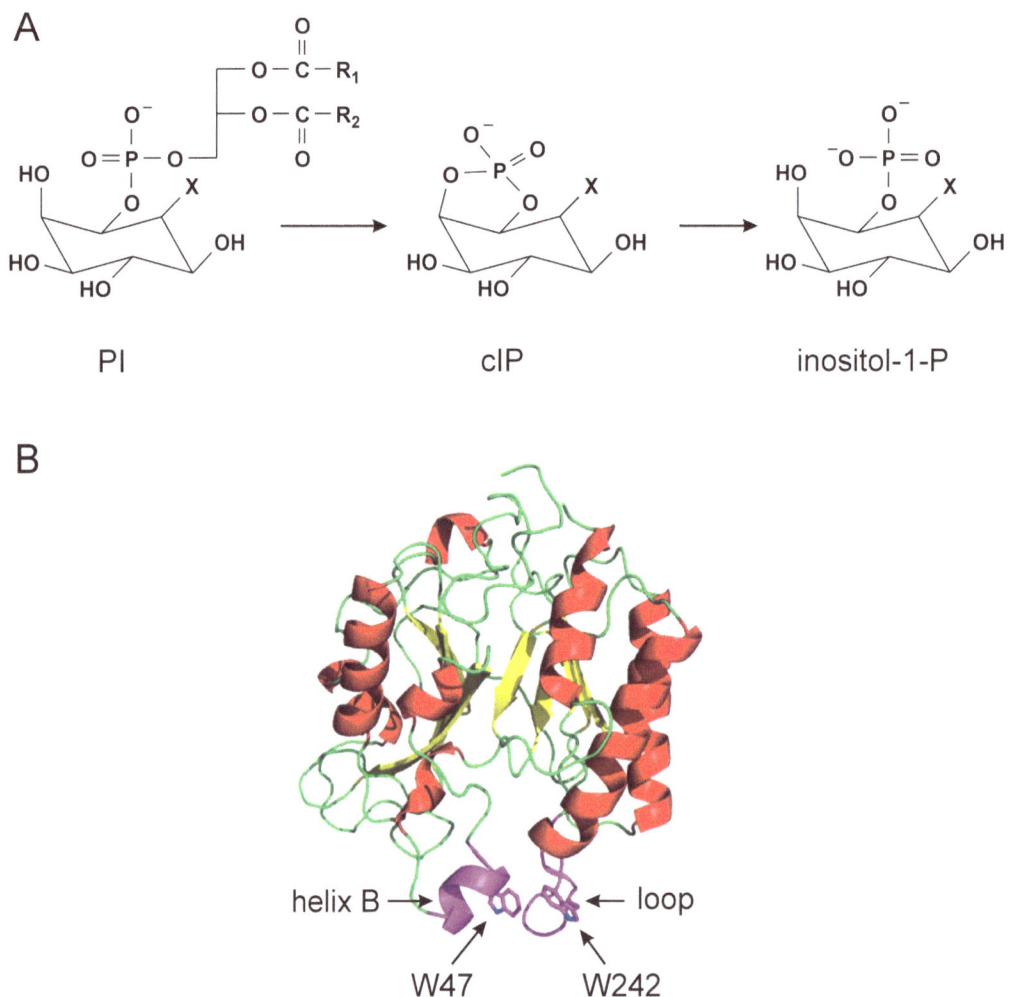

Figure 4: Bacterial PI-PLC enzymes. (A) The two-stage cleavage of PI/GPI-anchored proteins by bacterial PI-PLC enzymes. X indicates the position of the glycan chain of the GPI anchor. The first step produces diacylglycerol (not shown) and the cyclic inositol structure, cIP, thus forming the CRD, which can be detected by specific antibodies.

cIP is ring-opened in the second step to produce inositol-1-phosphate. (B) X-ray structural model of the PI-PLC from *B. cereus* in ribbon representation (generated from PDB 1PTD[79] using PyMOL). The location of helix B (residues 42-28) and the rim loop (residues 237-243) of the IBS are shown in magenta. Residues W47 and W242 are believed to insert into the outer leaflet of the lipid bilayer to facilitate association of the PI-PLC molecule with the membrane surface.

MODULATION OF PI-SPECIFIC PHOSPHOLIPASE CLEAVAGE BY MEMBRANE BIOPHYSICAL PROPERTIES

Although they are soluble enzymes, phospholipases are designed to work at surfaces, and they are generally more active when their substrates are presented at a micelle or bilayer surface, a phenomenon known as interfacial activation. Bacterial PI-PLCs display such effects in hydrolysis of both membrane-bound substrate (such as PI [104]), and the water-soluble intermediate cIP [105]. Cleavage takes place in two steps, with initial binding of the phospholipase to the interface mediated by the IBS, followed by catalysis. The binding process itself may take place in two stages; an initial electrostatic interaction, followed by stable binding mediated by hydrophobic interactions [82]. The physicochemical properties of the membrane itself may thus have a large modulatory effect on the cleavage of GPI anchors by PI-PLC. Examination of three different GPI-anchored enzymes in native plasma membrane vesicles and reconstituted bilayers showed that anchor cleavage by PI-PLCs from *B. thuringiensis* and *S. aureus* was modulated by electrostatic interactions between the enzyme and the membrane surface, and was greatly reduced in gel phase bilayers [6].

Later work using a reconstituted system containing the GPI-anchored enzyme, 5N-NTase, found that the rate of cleavage of the anchor by *B. thuringiensis* PI-PLC depended on the chain length and unsaturation of the membrane lipids [5]. Bilayers composed of highly fluid liquid crystalline lipids with low melting transitions gave very high rates of cleavage, whereas gel phase lipids cleaved the anchor ~1400-fold more slowly. A good correlation was found between the reduced temperature of the bilayer phase transition and the PI-PLC activity, suggesting that bilayer packing and fluidity are the most important modulators of activity. A temperature scan through the lipid bilayer phase transition showed a dramatic increase in PI-PLC activity (and a corresponding decrease in the activation energy of catalysis, Eact) at the melting transition. The aromatic and hydrophobic residues of the IBS helix/loop (especially Trp47 and Trp242) may be able to penetrate the membrane more easily, and thus stabilize PI-PLC on the surface, if the bilayer is highly fluid/more loosely packed, and thus more deformable. Another contributing factor may be a lower collision frequency of the PI-PLC enzyme with its membrane-bound substrate following stable interaction with the bilayer surface, caused by the very low rate of lateral diffusion in the gel state. In contrast, the incorporation of positively and negatively charged components (including gangliosides) to the bilayer had relatively small effects (2- to 4-fold) on the rate of PI-PLC cleavage [5]. These could arise in part from changes in lipid packing and fluidity, since these components also broadened the bilayer phase transition [5]. High PI-PLC activity was also observed for a GPI-anchored protein reconstituted into bilayers composed of raft-forming lipids (SM/cholesterol-rich liposomes) [5]. This may possibly be explained by local clustering of the GPI-anchored substrate in this lipid environment, which may mimic membrane microdomains, thus allowing each PI-PLC molecule to undergo more catalytic turnovers following a single membrane binding event.

More recent studies using PI as a substrate, rather than a GPI-anchored protein, found similar results, with reduced PI-PLC activity observed in bilayers with higher molecular order as judged by fluorescence polarization [106]. An interesting correlation was also observed between PI-PLC activity and the curvature of the liposomes, with activity decreasing dramatically as liposome diameter increased from 50-200 nm [106]. This is consistent with a previous report that smaller vesicles produced substantially higher PI-PLC activity than larger ones [82]. It seems likely that lipids in the highly curved outer leaflet of small vesicles occupy a larger surface area per molecule, thus allowing better penetration of the enzyme between their acyl chains.

REGULATION OF PROTEIN CONFORMATION AND FUNCTION BY THE GPI ANCHOR

The conformation of a GPI-anchored protein, and thus its function, may be altered by the presence of the anchor. Early in the history of GPI-anchored proteins, it was noted that 5N-NTase appeared to show increased activity after treatment with PI-PLC [107]. It was later confirmed that this protein was catalytically activated when the GPI anchor was removed by PI-PLC from both *B. thuringiensis* and *S. aureus* [6]. A detailed kinetic study of lymphocyte 5N-NTase recon-stituted into lipid bilayers showed that the KM for AMP remained essentially unchanged when the protein was converted into its soluble form, while Vmax was substantially increased [2]. The turnover number appeared to vary in the GPI-anchored form of the enzyme depending on the nature of the lipid bilayer, suggesting that membrane insertion of the anchor altered its catalytic efficiency. Similar effects were noted for 5N-NTase from chicken gizzard [108], and two forms of the liver enzyme (anchored and cleaved) also displayed different KM values [109]. These results suggested that the GPI anchor imposes conformational restraints on the enzyme (thus lowering its catalytic efficiency), which are relaxed following its removal by phospholipase cleavage. Large increases in catalytic activity after anchor cleavage were also reported for several other GPI-anchored enzymes, including the PH-20 hyaluronase [110], pig kidney dipeptidase [111], a metal-loproteinase from the trypanosomatid *Herpetomonas samuelpessoai* [112], and a *Plasmodium falciparum* protease [113]. Recent work in our lab has indicated that reconstituted PLAP also shows catalytic activation after release from the bilayer surface by PI-PLC (J.W.K. Chu and F.J. Sharom, unpublished data).

The conformation of non-enzyme proteins may also be affected by GPI anchor cleavage. When the anchor is removed from Thy-1 antigen by phospholipase C or D, a major conformational change takes place, resulting in greatly reduced binding of the OX-7 monoclonal antibody [68]. Later work found that several monoclonal antibodies were completely unable to bind to Thy-1 after anchor removal [114], suggesting that a substantial conformational change takes place. Similarly, GPI anchor cleavage from human CEA [115] and the very small (12 amino acid) human CD52 antigen [116] reduces binding of antibody. Antibody binding to GPI-anchored surface antigens in a *Paramecium* species [117] and in several *T. brucei* coat proteins [118] was also strongly influenced by the presence of the GPI anchor. Several other proteins, including a folate receptor variant [119], LFA-3 antigen [120], and yeast cAMP binding proteins [121], displayed reduced affinity for their substrates after GPI anchor removal, suggesting a change in protein tertiary structure. The widespread nature of these effects, ranging from microorganisms to humans, suggests that the GPI anchor may alter the conformation and function of all proteins to which it is linked. The effect of anchor removal on antigenicity has important consequences for production of vaccines against GPI-anchored protein targets. Recombinant proteins produced in bacteria, which are not GPI-anchored, may not be suitable as antigens because their conformation is not the same as the membrane-anchored version.

Changes in protein conformation induced by the presence of the GPI anchor may be transmitted directly from the anchor to the protein, or they may result from specific interactions of the protein with the membrane surface (see above), which are only possible in the GPI-anchored form. An FTIR study obtained evidence for a direct effect of the GPI anchor on the structure of bovine intestine AP [122]. The soluble form of the enzyme lacking the anchor displayed decreased stability of its secondary structure and conformation in response to thermal and pH denaturation. A recent study used nuclear magnetic resonance (NMR) spectroscopy and molecular dynamics simulations to explore the behaviour of a free GPI lipid in aqueous solution, and the same GPI lipid inserted into detergent micelles as a membrane mimic [123]. Results showed very little difference in the structure and dynamics of the pentasaccharide core of the anchor as a result of micelle insertion. Thus, it seems likely that the conformational changes imposed by the presence of the anchor are transmitted, in part, through the C-terminus of the protein itself.

MODULATION OF GPI-ANCHORED PROTEIN FUNCTION BY THE MEMBRANE

There is limited information available on the influence of membrane lipids and membrane physiochemical properties on the behaviour and activity of GPI-anchored proteins. An early study showed that the biophysical properties of the membrane affected the activity of 5N-NTase in rat intestine basolateral membranes [124]. A change in Eact was noted at a particular temperature, which coincided with a thermotropic phase transition in the membrane that could be detected by fluorescence polarization and DSC. Later work using purified 5N-NTase reconstituted into lipid bilayers reported a significant change in Eact at the phase transition temperature of the lipid bilayer [4]. At this time, 5N-NTase was believed to be an integral protein, so these results were not deemed to be unusual. However, they clearly demonstrate that the catalytic activity of this GPI-anchored protein is directly affected by the physical state of the membrane. We have recently shown that purified PLAP reconstituted into two different PC bilayers with defined melting points also shows different Eact values in the gel and liquid crystalline phases (J.W.K. Chu and F.J. Sharom, unpublished data). The effects of membrane fluidity are most likely mediated by contact between the GPI-anchored protein and the membrane surface, which might allow transmission of structural changes that would modify its catalytic properties. The function of GPI-anchored proteins, in general, may modulated by the phase state of the membrane.

Recent studies in model systems have shown that lipid packing, membrane order and bilayer curvature all appear to modulate the activity of PLAP [125]. Extruded vesicles containing reconstituted PLAP showed a monotonic decline in catalytic activity as the diameter of the proteoliposomes increased from 100 to 400 nm. This was the first report that the enzyme activity of a GPI-anchored protein is negatively affected by the lower bilayer curvature found in large vesicles. It was speculated that smaller, highly curved vesicles allow a larger spacing between proteins, thus enhancing catalytic turnover [125]. Similar consequences of high curvature have been noted for other amphitrophic enzymes that spend part of their time associated with the membrane surface, including protein kinase C [126], CTP:phosphocholine cytidylyltransferase [127] and PI-PLC [106]). This effect may be important in structures such as caveolae, where GPI-anchored proteins are sometimes located [128], since their "necks" contain highly curved regions that could enhance enzyme activity.

Rafts may inhibit the catalytic activity of GPI-anchored proteins located there. This was shown for PLAP, whose enzyme activity was inhibited by reconstitution into cholesterol-rich bilayers that mimicked the composition of lipid rafts [125]. The activity of a GPI-anchored phospholipase B1 from a fungal pathogen, which was clustered in rafts in the native membrane, was also down-regulated in this environment, and showed a 15-fold increase in activity following raft disruption with Triton X100 [129]. These effects may also be mediated through direct contact of the protein component with the bilayer surface.

REFERENCES

[1]   Reid-Taylor, K.L. et al. Reconstitution of the glycosylphosphatidylinositol-anchored protein Thy-1: interaction with membrane phospholipids and galactosylceramide. Biochem Cell Biol 77, 189-200 (1999).

[2]   Lehto, M.T. & Sharom, F.J. Release of the glycosylphosphatidylinositol-anchored enzyme ecto-5'-nucleotidase by phospholipase C: catalytic activation and modulation by the lipid bilayer. Biochem J 332, 101-109 (1998).

[3]   Sharom, F.J. & Eckford, P.D. Reconstitution of membrane transporters. Methods Mol Biol 227, 129-154 (2003).

[4]   Sharom, F.J. et al. Reconstitution of lymphocyte 5'-nucleotidase in lipid bilayers: behaviour and interaction with concanavalin A. Can J Biochem Cell Biol 63, 1049-1057 (1985).

[5]   Lehto, M.T. & Sharom, F.J. PI-specific phospholipase C cleavage of a reconstituted GPI-anchored protein: modulation by the lipid bilayer. Biochemistry 41, 1398-1408 (2002).

[6]   Sharom, F.J. et al. Modulation of the cleavage of glycosylphosphatidylinositol-anchored proteins by specific bacterial phospholipases. Biochem Cell Biol 74, 701-713 (1996).

[7]   Chan, P.Y. et al. Influence of receptor lateral mobility on adhesion strengthening between membranes containing LFA-3 and CD2. J Cell Biol 115, 245-255 (1991).

[8]   Lehto, M.T. & Sharom, F.J. Proximity of the protein moiety of a GPI-anchored protein to the membrane surface: a FRET study. Biochemistry 41, 8368-8376 (2002).

[9]   Chan, Y.H. & Boxer, S.G. Model membrane systems and their applications. Curr Opin Chem Biol 11, 581-587 (2007).

[10] Groves, J.T. & Dustin, M.L. Supported planar bilayers in studies on immune cell adhesion and communication. J Immunol Methods 278, 19-32 (2003).

[11] McConnell, H.M. et al. Supported planar membranes in studies of cell-cell recognition in the immune system. Biochim Biophys Acta 864, 95-106 (1986).

[12] Kloboucek, A. et al. Adhesion-induced receptor segregation and adhesion plaque formation: A model membrane study. Biophys J **77**, 2311-2328 (1999).

[13] Groves, J.T. et al. Electrical manipulation of glycan-phosphatidyl inositol-tethered proteins in planar supported bilayers. Biophys J 71, 2716-2723 (1996).

[14] Kihn, L. et al. Incorporation of human liver and placental alkaline phosphatases into liposomes and membranes is via phosphatidylinositol. Biochem Cell Biol 68, 1112-1118 (1990).

[15]   Kuchel, P.W. et al. Molecular weights of the Thy-1 glycoproteins from rat thymus and brain in the presence and absence of deoxycholate. Biochem J 169, 411-417 (1978).

[16]   Dustin, M.L. et al. Correlation of CD2 binding and functional properties of multimeric and monomeric lymphocyte function-associated antigen 3 J Exp Med 169, 503-517 (1989).

[17]   Schroeder, R. et al. Interactions between saturated acyl chains confer detergent resistance on lipids and glycosylphosphatidylinositol (GPI)-anchored proteins: GPI-anchored proteins in liposomes and cells show similar behavior. Proc Nat l Acad Sci U.S.A. 91, 12130-12134 (1994).

[18]   Schreier, H. et al. Targeting of liposomes to cells expressing CD4 using glycosylphosphatidylinositol-anchored gp120. Influence of liposome composition on intracellular trafficking. J Biol Chem 269, 9090-9098 (1994).

[19]   Camolezi, F.L. et al. Construction of an alkaline phosphatase-liposome system: a tool for biomineralization study. Int J Biochem Cell Biol 34, 1091-1101 (2002).

[20]   Medof, M.E. et al. Cell-surface engineering with GPI-anchored proteins. FASEB J 10, 574-586 (1996).

[21]   Rigaud, J.L. et al. Mechanisms of membrane protein insertion into liposomes during reconstitution procedures involving the use of detergents. 2. Incorporation of the light-driven proton pump bacteriorhodopsin. Biochemistry 27, 2677-2688 (1988).

[22]   Rigaud, J.L. et al. Reconstitution of membrane proteins into liposomes: application to energy-transducing membrane proteins. Biochim Biophys Acta 1231, 223-246 (1995).

[23]   Nosjean, O. & Roux, B. Ectoplasmic insertion of a glycosylphosphatidylinositol-anchored protein in glycosphingolipid- and cholesterol-containing phosphatidylcholine vesicles. Eur J Biochem 263, 865-870 (1999).

[24]   Nosjean, O. & Roux, B. Anchoring of glycosylphosphatidylinositol-proteins to liposomes. Methods Enzymol 372, 216-232 (2003).

[25]   Angrand, M. et al. Detergent-mediated reconstitution of a glycosyl-phosphatidylinositol-protein into liposomes. Eur J Biochem 250, 168-176 (1997).

[26]   Morandat, S. et al. Cholesterol-dependent insertion of glycosylphosphatidylinositol-anchored enzyme. Biochim Biophys Acta 1564, 473-478 (2002).

[27]   Colhone, M.C. et al. Incorporation of antigenic GPI-proteins from Leishmania amazonensis to membrane mimetic systems: influence of DPPC/cholesterol ratio. J Colloid Interface Sci 333, 373-379 (2009).

[28]   Nelson, K.L. et al. The glycosylphosphatidylinositol-anchored surface glycoprotein Thy-1 is a receptor for the channel-forming toxin aerolysin. J Biol Chem 272, 12170-12174 (1997).

[29]   Ronzon, F. et al. Insertion of a glycosylphosphatidylinositol-anchored enzyme into liposomes. J Membr Biol 197, 169-177 (2004).

[30]   Villar, A.V. et al. Towards the in vitro reconstitution of caveolae. Asymmetric incorporation of glycosylphosphatidylinositol (GPI) and gangliosides into liposomal membranes. FEBS Lett 457, 71-74 (1999).

[31]   Bonilla, J.B. et al. Phospholipase cleavage of D- and L-chiro-glycosylphosphoinositides asymmetrically incorporated into liposomal membranes. Chemistry 12, 1513-1528 (2006).

[32]   Brockman, H. Lipid monolayers: why use half a membrane to characterize protein-membrane interactions? Curr Opin Struct Biol 9, 438-443 (1999).

[33]   Ronzon, F. et al. Structure and orientation of a glycosylphosphatidyl inositol anchored protein at the air/water interface. J Phys Chem B 106, 3307-3315 (2002).

[34]   Ronzon, F. et al. A thermodynamic study of GPI-anchored and soluble form of alkaline phosphatase films at the air-water interface. *J.* Colloid Interface Sci 301, 493-502 (2006).

[35]   Caseli, L. et al. Adsorption of detergent-solubilized and phospholipase C-solubilized alkaline phosphatase at air/liquid interfaces. Colloids Surf B-Biointerfaces 30, 273-282 (2003).

[36]   Ronzon, F. et al. Behavior of a GPI-anchored protein in phospholipid monolayers at the air-water interface. Biochim Biophys Acta 1560, 1-13 (2002).

[37]   Caseli, L. et al. Surface density as a significant parameter for the enzymatic activity of two forms of alkaline phosphatase immobilized on phospholipid Langmuir-Blodgett films. J Colloid Interface Sci 275, 123-130 (2004).

[38]   Caseli, L. et al. Incorporation conditions guiding the aggregation of a glycosylphosphatidyl inositol (GPI)-anchored protein in Langmuir monolayers. Colloids Surf B Biointerfaces 46, 248-254 (2005).

[39]   Caseli, L. et al. Effect of molecular surface packing on the enzymatic activity modulation of an anchored protein on phospholipid Langmuir monolayers. Langmuir 21, 4090-4095 (2005).

[40] Kouzayha, A. & Besson, F. GPI-alkaline phosphatase insertion into phosphatidylcholine monolayers: phase behavior and morphology changes. Biochem Biophys Res Commun 333, 1315-1321 (2005).

[41] Giocondi, M.C. et al. Remodeling of ordered membrane domains by GPI-anchored intestinal alkaline phosphatase. Langmuir 23, 9358-9364 (2007).

[42] Frederix, P.L. et al. Atomic force microscopy of biological membranes. Biophys J 96, 329-338 (2009).

[43] Henderson, R.A. et al. Lipid rafts: Feeling is believing. News Physiol Sci 19, 39-43 (2004).

[44] Giocondi, M.C. et al. Characterizing the interactions between GPI-anchored alkaline phosphatases and membrane domains by AFM. Pflugers Arch 456, 179-188 (2008).

[45] Cross, B. et al. Measurement of the anchorage force between GPI-anchored alkaline phosphatase and supported membranes by AFM force spectroscopy. Langmuir 21, 5149-5153 (2005).

[46] Caseli, L. et al. Rat osseous plate alkaline phosphatase as Langmuir monolayer-an infrared study at the air-water interface. J Colloid Interface Sci 320, 476-482 (2008).

[47] Simons, K. & Ikonen, E. Functional rafts in cell membranes. Nature 387, 569-572 (1997).

[48] Cary, L.A. & Cooper, J.A. Signal transduction - Molecular switches in lipid rafts. Nature 404, 945-947 (2000).

[49] Paulick, M.G. & Bertozzi, C.R. The glycosylphosphatidylinositol anchor: a complex membrane-anchoring structure for proteins. Biochemistry (2008).

[50] Munro, S. Lipid rafts: Elusive or illusive? Cell 115, 377-388 (2003).

[51] Kahya, N. et al. Raft partitioning and dynamic behavior of human placental alkaline phosphatase in giant unilamellar vesicles. Biochemistry 44, 7479-7489 (2005).

[52] Morandat, S. et al. Role of GPI-anchored enzyme in liposome detergent-resistance. J. Membr. Biol. 191, 215-221 (2003).

[53] Milhiet, P.E. et al. AFM characterization of model rafts in supported bilayers. Single Molecules 2, 109-112 (2001).

[54] Saslowsky, D.E. et al. Placental alkaline phosphatase is efficiently targeted to rafts in supported lipid bilayers. J Biol Chem 277, 26966-26970 (2002).

[55] Giocondi, M.C. et al. Temperature-dependent localization of GPI-anchored intestinal alkaline phosphatase in model rafts. J Mol Recognit 20, 531-537 (2007).

[56] Noda, M. et al. High lateral mobility of endogenous and transfected alkaline phosphatase: a phosphatidylinositol-anchored membrane protein. J Cell Biol 105, 1671-1677 (1987).

[57] Phelps, B.M. et al. Restricted lateral diffusion of PH-20, a PI-anchored sperm membrane protein. Science 240, 1780-1782 (1988).

[58] Zhang, F. et al. Lateral diffusion of membrane-spanning and glycosylphosphatidylinositol-linked proteins: toward establishing rules governing the lateral mobility of membrane proteins. J Cell Biol 115, 75-84 (1991).

[59] Fein, M. et al. Lateral mobility of lipid analogues and GPI-anchored proteins in supported bilayers determined by fluorescent bead tracking. J Membr Biol 135, 83-92 (1993).

[60] Paulick, M.G. et al. Synthetic analogues of glycosylphosphatidylinositol-anchored proteins and their behavior in supported lipid bilayers. J Am Chem Soc. 129, 11543-11550 (2007).

[61] Paulick, M.G. et al. A chemical approach to unraveling the biological function of the glycosylphosphatidylinositol anchor. Proc Natl Acad Sci U.S.A. 104, 20332-20337 (2007).

[62] Sheets, E.D. et al. Transient confinement of a glycosylphosphatidylinositol-anchored protein in the plasma membrane. Biochemistry 36, 12449-12458 (1997).

[63] Vrljic, M. et al. Translational diffusion of individual class II MHC membrane proteins in cells. Biophys J. 83, 2681-2692 (2002).

[64] Pinaud, F. et al. Dynamic partitioning of a glycosyl-phosphatidylinositol-anchored protein in glycosphingolipid-rich microdomains imaged by single-quantum dot tracking. Traffic 10, 691-712 (2009).

[65] Subczynski, W.K. & Kusumi, A. Dynamics of raft molecules in the cell and artificial membranes: approaches by pulse EPR spin labeling and single molecule optical microscopy. Biochim Biophys Acta 1610, 231-243 (2003).

[66] Umemura, Y.M., Vrljic, M., Nishimura, S.Y., Fujiwara, T.K., Suzuki, K.G. & Kusumi, A. Both MHC class II and its GPI-anchored form undergo hop diffusion as observed by single-molecule tracking. Biophys J 95, 435-450 (2008).

[67] Homans, S.W. et al. Solution structure of the glycosylphosphatidylinositol membrane anchor glycan of Trypanosoma brucei variant surface glycoprotein. Biochemistry 28, 2881-2887 (1989).

[68] Barboni, E. et al. The glycophosphatidylinositol anchor affects the conformation of Thy-1 protein. J Cell Sci 108, 487-497 (1995).

[69] Perkins, S.J. et al. The Thy-1 glycoprotein: a three-dimensional model. Trends Biochem Sci 13, 302-303 (1988).

[70] Sharom, F.J. & Lehto, M.T. Glycosylphosphatidylinositol-anchored proteins: structure, function, and cleavage by phosphatidylinositol-specific phospholipase C. Biochem Cell Biol 80, 535-549 (2002).

[71] Griffith, O.H. & Ryan, M. Bacterial phosphatidylinositol-specific phospholipase C: structure, function, and interaction with lipids. Biochim Biophys Acta 1441, 237-254 (1999).

[72] Mengaud, J. et al. Identification of phosphatidylinositol-specific phospholipase C activity in Listeria monocytogenes: a novel type of virulence factor? Mol Microbiol 5, 367-372 (1991).

[73] Daugherty, S. & Low, M.G. Cloning, expression, and mutagenesis of phosphatidylinositol-specific phospholipase C from Staphylococcus aureus: a potential staphylococcal virulence factor. Infect Immun 61, 5078-5089 (1993).

[74] Ikezawa, H. Bacterial PIPLCs-unique properties and usefulness in studies on GPI anchors. Cell Biol Int Rep 15, 1115-1131 (1991).

[75] Broomfield, S.J. & Hooper, N.M. Characterization of an antibody to the cross-reacting determinant of the glycosyl-phosphatidylinositol anchor of human membrane dipeptidase. Biochim Biophys Acta 1145, 212-218 (1993).

[76] Roberts, W.L. et al. Lipid analysis of the glycoinositol phospholipid membrane anchor of human erythrocyte acetylcholinesterase. Palmitoylation of inositol results in resistance to phosphatidylinositol-specific phospholipase C. J Biol Chem 263, 18766-18775 (1988).

[77] Rosenberry, T.L. A chemical modification that makes glycoinositol phospholipids resistant to phospholipase C cleavage: fatty acid acylation of inositol. Cell Biol Int Rep 15, 1133-1150 (1991).

[78] Moser, J. et al. Crystal structure of the phosphatidylinositol-specific phospholipase C from the human pathogen Listeria monocytogenes. J Mol Biol 273, 269-282 (1997).

[79] Heinz, D.W. et al. Crystal structure of the phosphatidylinositol-specific phospholipase C from Bacillus cereus in complex with myo-inositol. EMBO J 14, 3855-3863 (1995).

[80] Heinz, D.W. et al. Crystal structure of phosphatidylinositol-specific phospholipase C from Bacillus cereus in complex with glucosaminyl(alpha 1-->6)-D-myo-inositol, an essential fragment of GPI anchors. Biochemistry 35, 9496-9504 (1996).

[81] Zhou, C. et al. Allosteric activation of phosphatidylinositol-specific phospholipase C: specific phospholipid binding anchors the enzyme to the interface. Biochemistry 36, 10089-10097 (1997).

[82] Wehbi, H. et al. Investigating the interfacial binding of bacterial phosphatidylinositol-specific phospholipase C Biochemistry 42, 9374-9382 (2003).

[83] Feng, J.W. et al. Role of tryptophan residues in interfacial binding of phosphatidylinositol-specific phospholipase C. J Biol Chem 277, 19867-19875 (2002).

[84] Jones, D.R. et al. Glycosyl-phosphatidylinositol-phospholipase type D: a possible candidate for the generation of second messengers. Biochem Biophys Res Commu 233, 432-437 (1997).

[85] Singh, N. et al. A novel class of cell surface glycolipids of mammalian cells. Free glycosyl phosphatidylinositols. J Biol Chem 271, 12879-12884 (1996).

[86] Sharom, F.J. & Radeva, G. GPI-anchored protein cleavage in the regulation of transmembrane signals. Subcell Biochem 37, 285-315 (2004).

[87] Davitz, M.A. et al. A glycan-phosphatidylinositol-specific phospholipase D in human serum. Science 238, 81-84 (1987).

[88] Low, M.G. & Prasad, A.R. A phospholipase D specific for the phosphatidylinositol anchor of cell-surface proteins is abundant in plasma. Proc Natl Acad Sci U.S.A. 85, 980-984 (1988).

[89] Hoener, M.C. et al. Isolation and characterization of a phosphatidylinositol-glycan-anchor-specific phospholipase D from bovine brain. Eur J Biochem 190, 593-601 (1990).

[90] Huang, K.S. et al. Purification and characterization of glycosyl-phosphatidylinositol-specific phospholipase D. J Biol Chem. 265, 17738-17745 (1990).

[91] Heller, M. et al. A novel form of glycosylphosphatidylinositol-anchor converting activity with a specificity of a phospholipase D in mammalian liver membranes. Biochim Biophys Acta 1109, 109-116 (1992).

[92] Hoener, M.C. & Brodbeck, U. Phosphatidylinositol-glycan-specific phospholipase D is an amphiphilic glycoprotein that in serum is associated with high-density lipoproteins. Eur J Biochem 206, 747-757 (1992).

[93] Deeg, M.A. & Davitz, M.A. Glycosylphosphatidylinositol-phospholipase D: a tool for glycosylphosphatidylinositol structural analysis. Methods Enzymol 250, 630-640 (1995).

[94]   Low, M.G. & Huang, K.S. Factors affecting the ability of glycosylphosphatidylinositol-specific phospholipase D to degrade the membrane anchors of cell surface proteins. Biochem J 279, 483-493 (1991).

[95]   Hari, T. et al. Uptake and intracellular stability of glycosylphosphatidylinositol-specific phospholipase D in neuroblastoma cells. Biochim Biophys Acta 1355, 293-302 (1997).

[96]   Tsujioka, H. et al. Posttranslational modification of glycosylphosphatidylinositol (GPI)-specific phospholipase D and its activity in cleavage of GPI anchors. Biochem Biophys Res Commun 251, 737-743 (1998).

[97]   Jones, D.R. et al. Phosphorylation of glycosyl-phosphatidylinositol by phosphatidylinositol 3-kinase changes its properties as a substrate for phospholipases. FEBS Lett 579, 59-65 (2005).

[98]   Vogel, M. et al. Soluble low-Km 5'-nucleotidase from electric-ray (Torpedo marmorata) electric organ and bovine cerebral cortex is derived from the glycosyl-phosphatidylinositol-anchored ectoenzyme by phospholipase C cleavage. Biochem J 284, 621-624 (1992).

[99]   Park, S.W. et al. Endogenous glycosylphosphatidylinositol-specific phospholipase C releases renal dipeptidase from kidney proximal tubules in vitro. Biochem J 353, 339-344 (2001).

[100] Park, S.W. et al. Glycosyl-phosphatidylinositol (GPI)-anchored renal dipeptidase is released by a phospholipase C in vivo. Kidney Blood Press Res 25, 7-12 (2002).

[101] Movahedi, S. & Hooper, N.M. Insulin stimulates the release of the glycosyl phosphatidylinositol- anchored membrane dipeptidase from 3T3-L1 adipocytes through the action of a phospholipase C Biochem J 326, 531-537 (1997).

[102] Kreuger, J. et al. Opposing activities of Dally-like glypican at high and low levels of Wingless morphogen activity. Dev Cell 7, 503-512 (2004).

[103] Traister, A. et al. Mammalian Notum induces the release of glypicans and other GPI-anchored proteins from the cell surface. Biochem J (2007).

[104] Volwerk, J.J. et al. Phosphatidylinositol-specific phospholipase C from Bacillus cereus at the lipid-water interface: interfacial binding, catalysis, and activation. Biochemistry 33, 3464-3474 (1994).

[105] Zhou, C. et al. Activation of phosphatidylinositol-specific phospholipase C toward inositol 1,2-(cyclic)-phosphate. Biochemistry 36, 347-355 (1997).

[106] Ahyayauch, H. et al. Modulation of PI-specific phospholipase C by membrane curvature and molecular order. Biochemistry 44, 11592-11600 (2005).

[107] Low, M.G. & Finean, J.B. Specific release of plasma membrane enzymes by a phosphatidylinositol-specific phospholipase C Biochim Biophys Acta 508, 565-570 (1978).

[108] Stochaj, U. et al. 5'-Nucleotidases of chicken gizzard and human pancreatic adenocarcinoma cells are anchored to the plasma membrane via a phosphatidylinositol-glycan. Biochem J 262, 33-40 (1989).

[109] Piec, G. & Le Hir, M. The soluble 'low-Km' 5'-nucleotidase of rat kidney represents solubilized ecto-5'-nucleotidase. Biochem J 273, 409-413 (1991).

[110] Gmachl, M. et al. The human sperm protein PH-20 has hyaluronidase activity. FEBS Lett 336, 545-548 (1993).

[111] Brewis, I.A. et al. Activation of the glycosyl-phosphatidylinositol-anchored membrane dipeptidase upon release from pig kidney membranes by phospholipase C. Biochem J 303, 633-638 (1994).

[112] dos Santos, A.L. et al. Activation of the glycosylphosphatidylinositol-anchored membrane proteinase upon release from Herpetomonas samuelpessoai by phospholipase C. Curr Microbiol 45, 293-298 (2002).

[113] Braun-Breton, C. et al. Induction of the proteolytic activity of a membrane protein in Plasmodium falciparum by phosphatidyl inositol-specific phospholipase C. Nature 332, 457-459 (1988).

[114] Kukulansky, T. et al. Cleavage of the glycosylphosphatidylinositol anchor affects the reactivity of Thy-1 with antibodies. J Immunol 162, 5993-5997 (1999).

[115] Durbin, H. et al. An epitope on carcinoembryonic antigen defined by the clinically relevant antibody PR1A3 Proc Natl Acad Sci U. S. A. 91, 4313-4317 (1994).

[116] Treumann, A. et al. Primary structure of CD52. J Biol Chem 270, 6088-6099 (1995).

[117] Capdeville, Y. et al. Allelic antigen and membrane-anchor epitopes of Paramecium primaurelia surface antigens. J Cell Sci 88, 553-562 (1987).

[118] Butikofer, P. et al. GPI-anchored proteins: now you see'em, now you don't. FASEB J 15, 545-548 (2001).

[119] Wang, X. et al. Variant GPI structure in relation to membrane-associated functions of a murine folate receptor. Biochemistry 35, 16305-16312 (1996).

[120] Tozeren, A. et al. Micromanipulation of adhesion of a Jurkat cell to a planar bilayer membrane containing lymphocyte function-associated antigen 3 molecules. J Cell Biol 116, 997-1006 (1992).

[121] Müller, G. & Bandlow, W. Lipolytic membrane release of two phosphatidylinositol-anchored cAMP receptor proteins in yeast alters their ligand-binding parameters. Arch Biochem Biophys 308, 504-514 (1994).

[122] Bortolato, M. et al. An infrared study of the thermal and pH stabilities of the GPI-alkaline phosphatase from bovine intestine. Biochem Biophys Res Commun 292, 874-879 (2002).

[123] Chevalier, F. et al. Structure and dynamics of the conserved protein GPI anchor core inserted into detergent micelles. Glycobiology 16, 969-980 (2006).

[124] Brasitus, T.A. & Schachter, D. Lipid dynamics and lipid-protein interactions in rat enterocyte basolateral and microvillus membranes. Biochemistry 19, 2763-2769 (1980).

[125] Sesana, S. et al. Membrane features and activity of GPI-anchored enzymes: alkaline phosphatase reconstituted in model membranes. Biochemistry 47, 5433-5440 (2008).

[126] Slater, S.J. et al. The modulation of protein kinase C activity by membrane lipid bilayer structure. J Biol Chem 269, 4866-4871 (1994).

[127] Cornell, R.B. Regulation of CTP:phosphocholine cytidylyltransferase by lipids. 1. Negative surface charge dependence for activation. Biochemistry 30, 5873-5880 (1991).

[128] Wang, J. et al. Evidence for segregation of heterologous GPI-anchored proteins into separate lipid rafts within the plasma membrane. J Membr Biol 189, 35-43 (2002).

[129] Siafakas, A.R. et al. Lipid rafts in *Cryptococcus neoformans* concentrate the virulence determinants phospholipase B1 and Cu/Zn superoxide dismutase. Eukaryot Cell 5, 488-498 (2006).

CHAPTER 4

Proteomic Approaches for GPI-Anchored Protein Analysis

Miren J. Omaetxebarria and Felix Elortza

Abstract; GPI-anchored membrane proteins (GPI-APs) are a functionally and structurally diverse protein family present in eukaryote cells. Their common feature is the anchoring mechanism via glycosylphosphatidylinositol (GPI) to a lipid located in the extracellular leaflet of plasma membrane. Following a modification specific proteomic strategy, GPI-APs can be selectively isolated by a combined treatment of membrane fractions with Triton X-114 detergent and phosphatidylinositol-specific phospholipase C and/or D. The released proteins can be further identified by liquid chromatography on-line coupled to a tandem mass spectrometer. Although the method is proven to be highly specific, some non-GPI-proteins are also known to be released. Bioinformatics has been regularly used to discriminate the GPI-APs among all the identified proteins with a high degree of accuracy. Structural characterization of the GPI-anchors has been historically a laborious analytical task. Recently, HILIC (Hydrophilic Interaction Liquid Chromatography) and titanium dioxide enrichment of GPI-modified peptides in combination with mass spectrometry have been used to isolate and analyse GPI-anchors as well as for the determination of the anchor attachment site. The amino acid where the previously synthesized GPI-anchor is attached is named the omega site (ω-site). Moreover, GPI-specific diagnostic ions detected in tandem mass spectra can potentially be used in large-scale proteomic experiments to track GPI-specific peptides in complex mixtures. All the information obtained by the mentioned strategies has been used for the development of an integrated computational and experimental proteomic approach designed for identification of GPI-anchored peptides in MS/MS spectra as well as for ω-site determination.

INTRODUCTION

Post-translational modifications (PTMs) modulate the activities and functions of proteins. Thus, the study and analysis of specific PTMs is of paramount importance in order to get insight into their role in protein dynamics within cellular processes. The so called "modification specific approach" is often applied to the study of PTMs [1]. Combination of traditional biochemical protocols and molecular cell biology procedures together with mass spectrometry and bioinformatic tools makes the analysis of PTMs a feasible, yet challenging, task.

Glycosylphosphatidylinositol-anchored proteins (GPI-APs) are a subset of post-translationally modified proteins, and it is estimated that 0.5 % of the genes in eukaryotic cells encode GPI-APs [2]. The post-translational processing of the pro-protein synthesized on the ribosome is a complex and metabolically expensive process which is completed in two main steps: first, the pro-protein is translocated to the endoplasmic reticulum, where after proteolytic processing of the C-terminal domain, a transamidase attaches *en bloc* the already formed GPI-anchor to this new C-terminus termed the ω-site [3]. The general structure of the anchor consists of a lipid tail attached to a glycan core. The addition of the GPI-anchor tags the protein so that it is tethered to the cell's extracellular leaflet in the plasma membrane. Unlike proteins modified by the addition of myristoil, palmitoyl and prenyl groups, GPI-modified proteins are the only ones sitting on the extracellular side of the cell and the big majority of them lack transmembrane spans. Considering their localization it appears obvious that the functions that have been attributed to GPI-APs are in the class of surface receptors, cell adhesion molecules, cell surface hydrolases or coat proteins, to mention some. Inversely, knowledge of the presence of a GPI-anchor may provide valuable information on the localization and functions of a particular protein.

The information for a gene product to be prone to be modified by the addition of a GPI-anchor is in the protein sequence itself, as certain general features are present in all known GPI-anchored pro-proteins (Fig. **1**): (i) a cleavable N-terminal hydrophobic signal peptide for translocation across the endoplasmic reticulum, (ii) absence of transmembrane domains between the signal peptide and the ω-site, (iii) a cleavable hydrophobic region in the COOH-terminus and (iv) a hydrophilic spacer region preceding the COOH-terminal hydrophobic region [4]. All these features are taken into account for the amino acid sequence analysis performed by existing bioinformatic tools used for the prediction of GPI-APs. A number of predictors are currently publicly available on-line, including big-π by Einsenhaber *et al.* at the University of Vienna [5] (http://mendel.imp.ac.at/gpi/gpi_server.html), DGPI by Kronegg and Buloz from the University of Geneva (retrieved from http://www.expasy.ch/tools/), GPI-SOM by Fankhauser and Mäser [6] from the University of Bern (http://gpi.unibe.ch/), FragAnchor algorithm by Poisson *et al.* [7] (http://navet.ics.hawaii.edu/~fraganchor/NNHMM/NNHMM.html) and the recently reported PredGPI by Pierleoni *et al.* [8] (http://gpcr.biocomp.unibo.it/predgpi/pred.htm). Although these bioinformatics tools are very useful, their predicting performance is compromised by the small size of the experimentally verified GPI-APs data

John A. Dangerfield & Christoph Metzner (Eds)

sets; as an example, is worth noting that only 10% of the proteins that are annotated as GPI-APs in the currently available protein databases have their ω-site experimentally verified. Therefore, a detailed, large-scale analysis of these modified proteins is required.

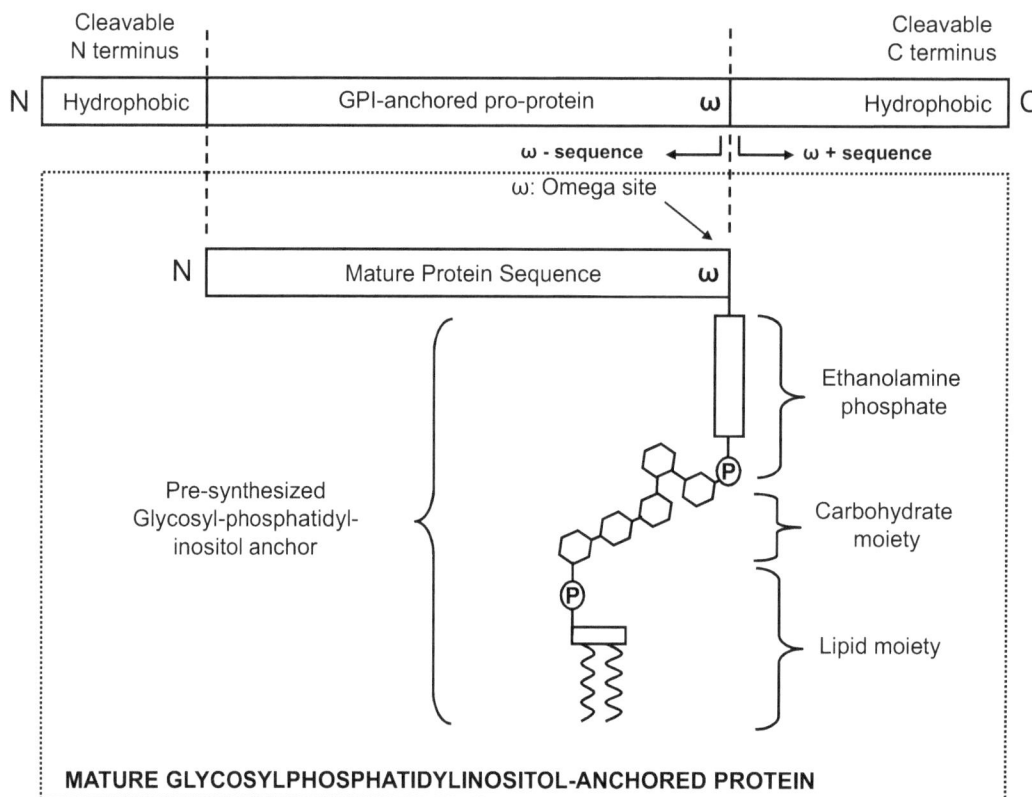

Figure 1: Schematic view of the GPI-anchored pro-protein and mature protein.

Due to the biological relevance and the increasing interest in biomedicine of this group of proteins, the development of methods that will promote the understanding of their molecular details is mandatory. However, the difficulty to generally analyze GPI-anchored proteins and the GPI anchors hinders the achievement. The complex structure and the amphiphilic nature of the anchor hamper the extraction, purification and fully characterization of these molecules. Traditional biochemical approaches have been limited to single protein analysis, demanding in all cases large amounts of purified proteins.

Modification specific proteomics approaches dedicated to the analysis of GPI-APs have been reported and proven to be of great use. Mass spectrometry (MS) is a very valuable tool when working in proteomics; the sensitivity and accuracy of modern mass spectrometers allow the analysis of nanogram levels of individual proteins (low femtomol), while in the case of analysis of complex samples μg range of starting material is required (picomol).

This chapter covers the application of several proteomic strategies for the analysis and identification of GPI-APs in complex samples as well as for the detailed characterization of the structure of the glycan moiety of the anchor in individual pure proteins.

MODIFICATION SPECIFIC PROTEOMIC STRATEGY FOR GPI-AP ANALYSIS

Isolation of GPI-APs From Human Lipid Raft-Enriched Fractions and Plant Membranes

Proteomics is a rapidly expanding field that aims the systematic study of protein structure, function, interactions and dynamics. In general, the analysis of post-translationally modified proteins faces several challenges. A general complication of proteomic analysing is the wide dynamic range of protein expression within the cell. If the proteins of interest happen to be in low copy number per cell, a whole cell extract analysis will have few chances of success. Therefore, different sample preparations for specific protein enrichment have been developed. One way to tackle the problem is to follow the so called organellar subproteomic analysis, where the organelle in which the protein of interest is present is highly enriched or purified. In our case, it is known that GPI-APs are located in plasma membranes. Besides, due to the physicochemical characteristics of the lipid moiety of the anchor, they are known to be enriched in rafts, sphingolipid and cholesterol enriched domains located in the external leaflet of the plasma membrane [9-11]. Thus, the first step for a GPI-anchored targeted analysis is the purification of the plasma membrane and/or the lipid rafts. Obviously, GPI-APs are not the only class of proteins present in plasma membrane and rafts, so further purification steps considering the specific characteristics of the GPI-anchoring post-translational modification are needed.

Although GPI anchoring is an irreversible process, enzymes such as phosphoinositol-phospholipase C (PI-PLC) [12] and phospholipase D (PLD) [13,14] can specifically hydrolyse the anchoring moiety; this feature makes these enzymes extremely attractive for GPI-anchored proteins isolation. The enzymatic activity of these phospholipases in combination with non-ionic detergents such as Triton X-114 [15] allows a specific separation of GPI-APs from plants' plasma membranes [16-18] and human cell raft-enriched fractions [19] (Fig. **2**). Once the membranes have been purified, the protocol for GPI-AP isolation is rather straight forward (Protocol 3).

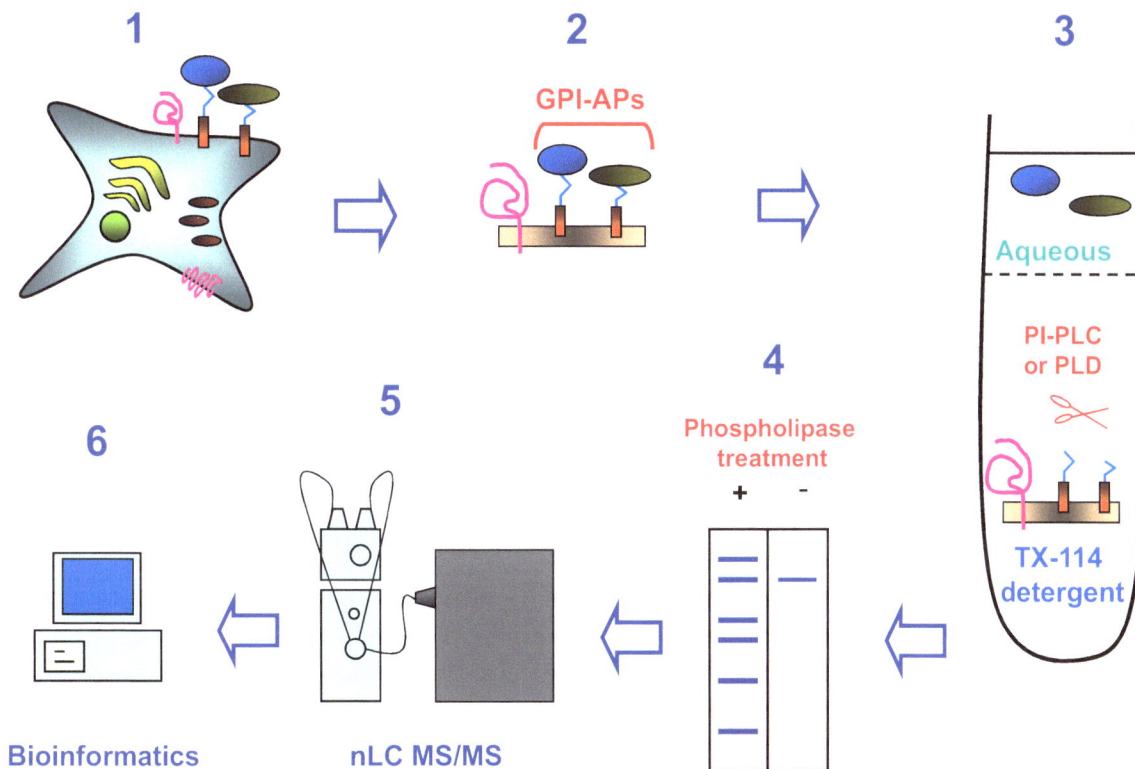

Figure 2: Modification specific proteomic strategy for GPI-AP analysis. Clockwise: 1) Eukaryotic cells. 2) Plasma membranes or rafts isolated from cells where the GPI-APs are located. 3) GPI isolation by combination of GPI specific phospholipases (Phosphoinositol specific Phospholipase C and D) and two phase separation induced by Triton X-114 detergent. 4) Isolated proteins are concentrated (*e.g.* by precipitation) and resolved by gel electrophoresis. Following gel staining, sliced protein bands are in-gel digested. 5) Resulting peptides are analysed by nano scale liquid chromatography coupled on-line with a tandem mass spectrometer (nLC-MS/MS). 6) Bioinformatic tools are used for protein GPI-AP identification and characterization.

****

PROTOCOL 3

Isolation of GPI-AP by phospholipase treatment in combination with Triton X-114 induced two-phase separation:

1. *25 µl of plant of human cell derived membranes are equilibrated by resuspending the pellet in buffer A (20 mM HEPES pH 7.5). Phenylmethylsulfonyl fluoride (PMSF) HEPES pH 7.5, and protease inhibitor are added to buffer A.*

2. *Spin down the suspension by centrifuging at 20,000 x g for 20 min at 4 °C.*

3.   *The membrane fraction is resuspended in 100 μl of buffer A, and then same volume of chilled Triton X-114 is added.*

4.   *The mixture is mixed to homogeneity.*

5.   *The mixture is chilled on ice for 5 min and then incubated at 37°C for 20 min for phase separation. The two phases will be formed, being the detergent phase on the bottom and the aqueous one on top of it.*

6.   *Washing step: the aqueous upper supernatant where membrane associated, but not intrinsic, proteins are released to is discarded (this procedure can be repeated for a further washing step).*

7.   *The detergent phase is recovered and 100 μl of buffer A with 2 units of phosphoinositol-phospholipase C (PI-PLC) (Molecular Probes Inc., Eugene, OR) or phospholipase D (PLD) is added.*

8.   *The mixture is incubated at 37°C gently shaking for 1 hr.*

9.   *Phase separation procedure is performed (step 5) and the GPI-AP containing aqueous supernatant is recovered for further analysis.*

****

GPI-Anchored Protein Identification: Mass Spectrometry and Bioinformatic Analysis

Mass spectrometry is a very valuable technique in protein chemistry and also in proteomics [20]. A mass spectrometer determines the mass to charge ratio (m/z) of peptides and protein ions. The separation of peptides by reversed-phase capillary high-performance liquid chromatography coupled on-line by electrospray ionisation to tandem mass spectrometers is a widely used methodology for complex protein-peptide mixture analysis. Proteins are identified relying on the automated, mass specific selection and collision-induced dissociation of peptide ions inside the mass spectrometer [21]. The combination of these methods with the tailored bioinformatic tools called searching engines, emerges as a routine analytical approach for protein identification.

Proteins recovered in the GPI-AP enriched aqueous supernatant can be concentrated by protein precipitation. Then, different procedures may be performed for final protein identification. Proteins can be resolved by sodium dodecyl sulphate polyacrylamide gel electrophoresis (*e.g.* SDS-PAGE). Following gel staining, sliced protein bands can be in-gel digested for further nano scale liquid chromatography tandem mass spectrometry analysis (nLC MS/MS). Alternatively, precipitated proteins may be resuspended in an appropriate buffer in order to perform in solution digestion. Recovered peptides can be equally analyzed by nLC MS/MS (Fig. **2**).

Although the method explained in Protocol 3 has proven to be rather specific for GPI-APs isolation, a few membrane-released proteins might not be GPI-APs. Hence, the identified proteins need to be validated as GPI-APs. An easy and straight forward method to know whether a protein is GPI-anchored or not, is to analyze the amino acid sequence of the gene product or pro-protein. As explained above in this chapter, the reason behind this fact is that all known GPI-APs although displaying different functionalities, share some common sequence related features (Fig. **1**): a hydrophobic N-terminal secretion signal that targets the protein to the endoplasmic reticulum and a C-terminal GPI signal anchor sequence. The amino acid residue to which the GPI is attached to is referred to as the "ω-site". Adjacent amino acids towards N-terminal from the "ω-site" are termed ω minus (ω -) and those amino acids towards C-terminal are termed ω plus (ω +). The general features of an GPI anchoring signal sequence (GSS) are: i) a stretch of ~10 polar amino acids (ω -10 to ω -1) that form a flexible linker region; ii) the ω amino acid, typically a small amino acid such as G, A, S, N, D, or C; iii) the ω +2 amino acid, the most restrictive position, predominantly G, A, or S; iv) a spacer region of moderately polar amino acids (ω +3 to ω +9 or more); and v) a stretch of hydrophobic amino acids variable in length but capable of spanning the membrane. A given C-terminal sequence may contain more than one set of candidates for the ω and ω +2 amino acids. Slight variations have been reported in the relative lengths of the spacer regions and hydrophobic segments of metazoan and protozoan signal anchor sequences [22-24]

Be all that as it may, publicly available predictors still show discrepancies for sequence analysis-based GPI-anchoring determination and the disagreement is even more noticeable when ascertaining the ω-site (Table **1**). Eleven proteins in a human cell raft-enriched fraction were identified and determined to be GPI-APs by Elortza and co-workers. However, it needs to be mentioned that not all the predictors used in the experiment succeeded in recognizing them as such [19]; moreover, the ω-sites assigned to the identified GPI-APs were not ascertain to the same residue by the different predictors used in the experiment. There are several reasons that may explain this lack of accuracy in the predictions. On one hand, it is obvious that the machine learning schemes are heavily dependent on the training set. It is not only the limitation of experimental results that makes the design of predictors problematic but also the annotation and curation of the scientific papers describing the results should be considered with caution. In Swiss-Prot, the protein sequence sites are annotated as experimentally verified, potential, probable or by similarity. In theory, one should only use the experimentally verified sites as a learning set for a predictor; however, in practise similarity is often used. Another issue is that it is not so straight forward to verify the quality of the experimental data in Swiss-Prot. As it is shown in Table 1, even for proteins whose ω-sites have been experimentally determined (*e.g.* CD59 or Folate Receptor) the different predictors do not show the same outcome in their analysis. Having said this, attention should be paid to the fact that one thing is the lack of accuracy of the predictors mentioned in this chapter when assigning the ω-site of a GPI-AP, and another one is the GPI-anchor existence prediction, which is in any case far more accurate [5,8].

Table 1: Omega site determination by different GPI anchoring predictors.

Swiss-Prot	GPI-anchored protein	Omega-site position given by predictors				
		Big-Pi	DGPI	GPI-SOM	Frag-Anchor	Pred-GPI
P05186	Alkaline phosphatase	502	NO/501	500	498	501
P08174	Decay acceleration factor (CD55)	353	353	NO/349	349	353
Q9BR17	Mesothelin/ megakaryocyte potentiating factor	606	606	625	606	606
Q7Z3B1	Neuronal growth regulator 1	NO/324	324	321	NO	325
P15328	Folate receptor	234	234	229	234	234
P14384	Carboxypeptidase M precursor	NO/421	423	422	423	418
P55290	Cadherin 13 preproprotein	693	690	689	692	690
P19256-2	Antigen CD58/ Surface glycoprotein LFA-3	NO/208	208	220	NO	219
P13987	CD59	102	102	120	102	102
Q03405	UPAR	305	304	304	304	304
Q10589	Bone marrow stromal cell antigen (CD317)	161	160	160	161	157

GPI-anchored protein: Swiss Prot database entry and protein name. Omega-site position given by predictors:
Big-Pi: Eisenhaber *et al.* [5] ; http://mendel.imp.ac.at/sat/gpi/gpi_server.html
DGPI: Kronegg and Buloz; http://www.expasy.org/tools/
GPI-SOM: Fankhauser and Mäner [6]; http://gpi.unibe.ch/
FragAnchor: Poisson *et al.*[7]; http://navet.ics.hawaii.edu/~fraganchor/NNHMM/NNHMM.html
PredGPI: Pierleoni *et al.* [8]; http://gpcr.biocomp.unibo.it/predgpi/pred.htm

Therefore, although bioinformatic analysis provides scientific community with a fast and reliable way to discriminate GPI-APs among other membrane proteins, the discrepancies between predictors reveals the need for further developing and tuning of the existing algorithms. As explained above, a limiting factor in the process for further refining of the predictors is the lack of reasonable experimental data on the characterization of the ω-sites of experimentally verified GPI-APs. Thus, biochemical verification of putative GPI modifications is still demanded. In this direction, western blotting analysis using a CDR antigen-specific antibody can be contemplated in order to confirm whether a PI-PLC released putative protein is GPI-anchored or not. The CDR antigen is the epitope formed in the GPI-AP sugar moiety located opposite to the amino termini. However, this approach is restricted to PI-PLC released putative GPI-APs and unspecific interaction cannot be ruled out; moreover, no additional information regarding the ω-site is provided [4]. Thus, technical and methodological developments for unequivocal GPI-anchoring and ω-site determination are needed. Another turn of force in the analysis of the GPI-APs is the characterisation of the amphipatic anchor. Although the general structure of the anchors seem to be well conserved among different species, their characterization is not a trivial task, due to, among other reasons, the various modifications found for the carbohydrate and lipidic moieties of the anchor. Even if the significance of these anchor variants is not known yet, the structural studies should not be neglected in order to fully understand the implications of the GPI modifications in biological processes.

ISOLATION AND CHARACTERIZATION OF THE GLYCAN MOIETY OF GPI ANCHORS: COMBINATION OF MINIATURIZED ENRICHMENT TECHNIQUES AND MASS SPECTROMETRY

Due to the nature of GPI-APs, a structure composed by protein and carbohydrate and lipid moieties, the structural characterization of the likes is not trivial. The complete elucidation of the GPI-anchor requires its release from the protein and the determination of the sugar and lipid moieties as well as of the existing heterogeneities on the presumably well conserved structure.

Several strategies based on combinations of diverse analytical methods have been reported for GPI-anchor structural analysis including enzymatic and chemical degradation of the isolated anchor and composition-, sequence- and linkage-analysis by gas or liquid chromatography and mass spectrometry (GC-MS or LC-MS). The GPI-anchor structures of *Trypanosoma cruzi* [25], human and bovine acetylcholinesterase [26,27], yeast glycoprotein [28], *Dictyostelium discoideum* prespore-specific antigen [28] and porcine and human membrane dipeptidase (MDP) [29], among others, have been described. Those studies suggest that the structure of the anchors is well-conserved among species. The general structure of a GPI-anchor consists of a glycerol or ceramide lipid moiety, and an inositol ring linked to a glycan core, composed by a glucosamine and three mannose residues, linked to an ethanolamine by a phosphodiester bond which in turn is linked to the ω-site of the protein via an amide bond (Fig. **3**). The presence of the nonacetylated glucosamine (GlcN) is a unique feature of GPI-anchors. Diversification of the core glycan structure is common due to the addition of extra side chains, generally glycans (i.e. HexNAc) or EtNP residues, which may be attached to the anchor. Furthermore, acylation of the inositol residue may occur and this variation may be responsible for the resistance of some GPI-APs to PI-PLC digestion, resistance that can be overcome by using PLD instead.

Mass spectrometry-based strategies are extensively used for post-translational modification studies. Nevertheless, there are still limitations regarding the detection of modified peptides due to their physicochemical properties (size, charge, hydrophobicity and/or lability) and mass spectrometric analysis related problems such as ionisation efficiency related ones. With the aim of overcoming those problems PTM specific enrichment strategies have been proposed, and proven very promising; immobilized metal affinity chromatography (IMAC), titanium dioxide (TiO$_2$), graphite powder, and hydrophilic interaction chromatography (HILIC), proved very useful for the recovery of post-translationally modified peptides. Such is the case for GPI-modified peptides; isolation of the GPI-modified peptides facilitates their characterization and this is achieved by using miniaturized enrichment techniques in the form of hydrophilic interaction liquid chromatography (HILIC) or titanium dioxide (TiO$_2$) microcolumns.

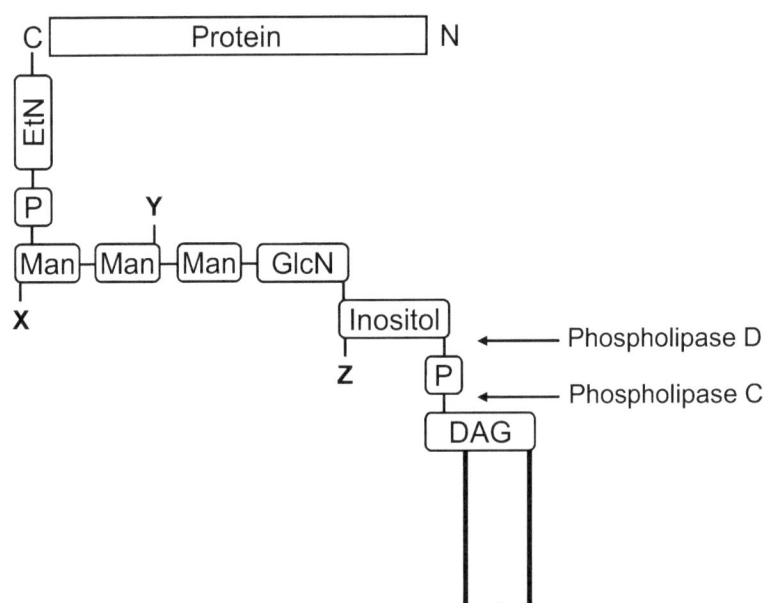

Figure 3: General structure of the GPI-anchor. Schematic general structure of a GPI-anchor. DAG: lipid moiety; Inositol: inositol ring; GlcN: glucosamine; Man: mannose residue; EtN: ethanolamine: P: phosphate group. X, Y and Z show positions in which the anchor may show variability. C and N show the carboxyl and amino terminal of the GPI-anchored protein.

HILIC and TiO₂ Enrichment of GPI-Modified Peptides: Analysis by Mass Spectrometry

HILIC is broadly used for the separation of a large range of compounds, such as carbohydrates, oligonucleotides, amino acids, metabolites and peptides. In HILIC, a polar stationary phase is used in combination with a less polar mobile phase. HILIC has also proven very valuable in proteomics, reducing the complexity of peptide/glycopeptide mixtures by depleting hydrophobic peptides and retaining the hydrophilic ones. This technique allows enrichment and targeted analysis of PTMs such as glycosylation [30] or N-acetylation [31] among others.

In the case of GPI-APs, phospholipase treated GPI-modified peptides are successfully captured by the use of HILIC, following the protocol described in Protocol 4 [32].

****

Protocol 4

Phospholipase Released GPI- Modified Peptide Enrichment by HILIC

1. *SDS-PAGE resolved protein band is sliced and in-gel digested. The resulting peptide mixture is dried in a vacuum centrifuge.*

2. *HILIC microcolumn preparation: HILIC material is packed in a partially constricted GeLoader tip (Eppendorf) to a column height of 5 mm.*

3. *The column is equilibrated in 80% acetonitrile (ACN)/0.5% formic acid (FA).*

4. *The dry peptide mixture is redissolved in 5 μl of 80% ACN/0.5% FA and loaded onto the HILIC microcolumn using air pressure.*

5. *Wash the HILIC microcolumn three times with 20 μl 80% ACN/0.5% FA.*

6. *Elute the bound peptides with 2 μl of 95% H₂O/5% FA, collecting them in a microcentrifuge tube.*

7. *Analysis of enriched fractions is performed by MALDI- or ESI- MS and/or MS/MS*

****

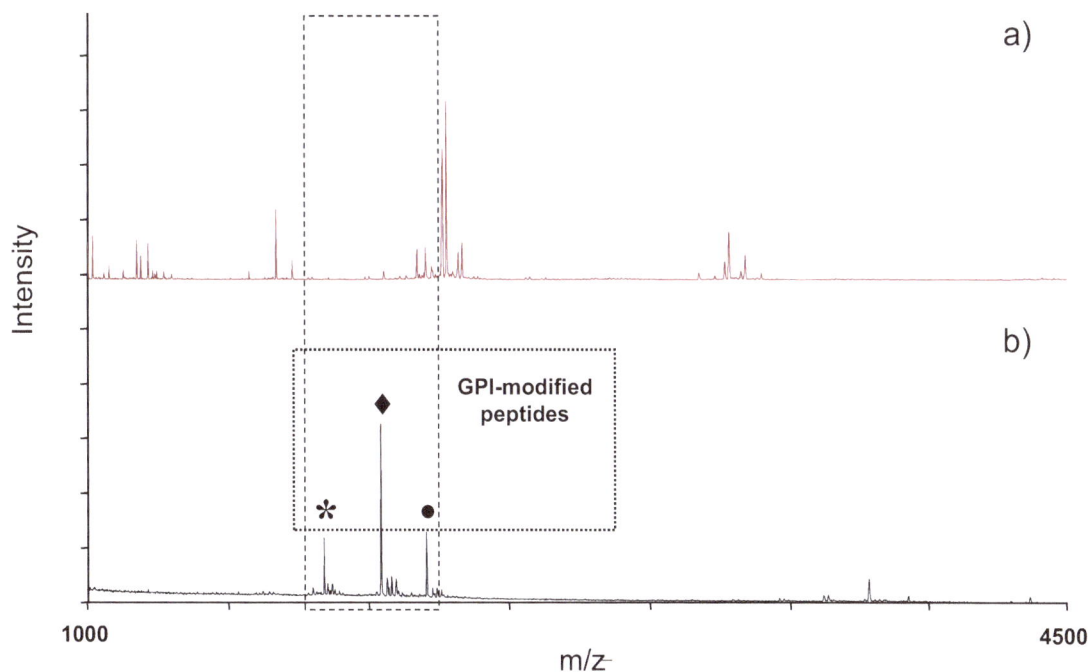

Figure 4: GPI-modified peptide purification by hydrophilic interaction chromatography (HILIC). MALDI-TOF MS spectra of 1 pmol of trypsin-digested and phospholipase released MDP, after purification on either a R2 reverse-phase microcolumn (a) or a HILIC microcolumn (b). Highlighted the GPI-modified peptides. (*) GPI-modified peptide of MDP, (♦) variant corresponding to core + HexNAc and (●) variant corresponding to core + HexHexNAc.

Mass spectrometry measurements give accurate mass to charge ratio (m/z) values of biomolecular ions generated by matrix assisted laser desorption ionisation (MALDI) or electrospray ionisation (ESI), and can be used for accurate molecular weight determination of phospholipase-released GPI-anchors. The outcome of the methodology described herein can be observed for the phospholipase released porcine membrane dipeptidase (MDP) GPI-AP (Fig. **4**), where successful GPI-modified peptide enrichment is achieved. The HILIC enriched fraction was in this case analyzed by MALDI-TOF MS. The peptide ion signal corresponding to the C-terminal tryptic GPI-modified peptide carrying the conserved structure EtNP-6Manα1-2Manα1-6Manα1-4GlcNα1-6InsP (*) is detected for the HILIC-purified sample at m/z 1839.6, which matches the calculated mass of the core glycan structure of the modified peptide of MDP. Together with this, signals corresponding to variants of the MDP anchor reported as core + HexNAc (♦) and core + HexHexNAc (•) were also detected (b). No signal corresponding to the GPI-modified peptides is detected in the non-enriched fraction (a).

TiO$_2$ chromatography is recognised as an established procedure for phosphopeptide enrichment [33]. Taking into account the structural composition of the GPI-anchors and the strong adsorption of phosphate-containing compounds to this material it can also be used for GPI-anchored peptide purification. Following standard TiO$_2$ phosphopeptide enrichment procedure the modified peptides of MDP can be effectively purified rendering basically the same result than the HILIC material [34].

No non-modified peptides are detected either in the HILIC nor in the TiO$_2$ enriched fractions suggesting the high specificity of these materials for GPI-modified peptides. Both enrichment methods are compatible with standard biochemical methods and allow analysis of low-picomole levels of GPI-APs. Application of such enrichment methods in shot-gun GPI-AP studies seems plausible.

Glycan Moiety Structure Analysis of GPI-Anchors by Tandem Mass Spectrometry

The use of tandem mass spectrometry (MS/MS) enables sequencing, characterization and identification of peptides, glycans and lipids. Consequently, MS/MS analysis provides accurate molecular weight values of the GPI-anchor and its fragments. Ions specifically generated as a consequence of fragmentation of modified peptides are called diagnostic ions and are very useful in the detection and identification of modifications. That is the case for GPI-peptides; MS/MS fragmentation of the HILIC- or TiO$_2$-captured GPI-modified peptides can be used for the study of the modified peptide glycan structure. The MS/MS fragmentation pattern of the glycan moiety of GPI-anchors occurs in a sequential manner (Fig. **5**) and interestingly enough, renders GPI-specific diagnostic ions. The exclusive presence of non-acetylated glucosamine (GlcN) in GPI-anchors and its diagnostic ion detection in MS/MS experiments (m/z 162.1) provides substantial information. In addition to GlcN, diagnostic ions corresponding to fragments mannose ethanolamine phosphate (ManEtNP) at m/z 286.1, glucosamine inositol phosphate (GlcNInsP) at m/z 422.1, and mannose ethanolamine phosphate glucosamine (ManEtNPGlcN) at m/z 447.2, are readily detected (Fig. **5**). The detection of these diagnostic ions can be considered as a strong indication for the presence of a GPI-anchor and has already proved to be of great use in modification specific proteomic studies when tracking GPI peptides in complex mixtures.

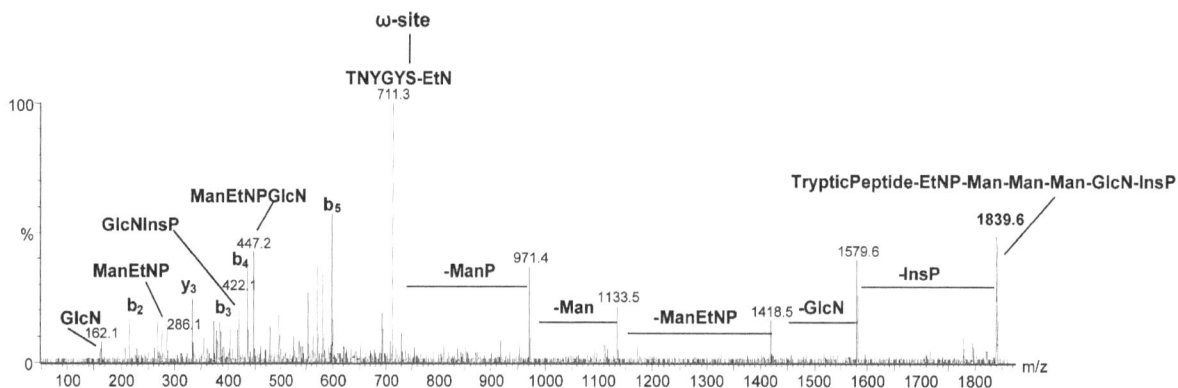

Figure 5: Tandem mass spectrometry analysis of GPI-modified peptide. MALDI-QTOF-MS/MS spectra of the core GPI-modified peptide at m/z 1839.6. Fragments and diagnostic ions corresponding to the tryptic C-terminal GPI-modified peptide of phospholipase released MDP are indicated.

GPI ANCHORING SITE DETERMINATION IN PROTEOMICS EXPERIMENTS

The questionable performance of the existing algorithms devoted to GPI-AP prediction and ω-site prediction has been mentioned earlier in the chapter. Among some other reasons, the lack of integration of prediction algorithms with experimental results appears important in this matter. Hence, results point to think that the data (i.e. GPI diagnostic ions, sequential fragmentation of the sugar moiety) obtained by mass spectrometric techniques may be beneficial for GPI-AP identification and characterization studies. The majority of the existing GPI-AP prediction algorithms rely solely on the combined information obtained from a signal predictor, a C-terminal transmembrane predictor and a predictor that works on the amino acid sequences surrounding the ω-site of previously reported GPI-APs. A new tool named VEMS-GPI (Virtual Expert Mass Spectrometrist for GPI analysis), that integrates computational and experimental proteomics approaches together with GPI-specific LC-MS/MS data is now available [35]. The outline of the algorithm is depicted in Fig. **6**. Five different steps are followed before a final conclusion on GPI-AP identification is drawn. Following a standard database dependent search of the MS/MS spectra of the sample under study, where GPI-APs together with contaminants are present, a prediction of GPI-APs and their corresponding possible ω-sites is made. The ten most likely ω-sites for each GPI-AP are considered for generating a *in silico* ω-site bearing C-terminal semitryptic peptide index, whose theoretical MS/MS spectra will be compared to the observed MS/MS spectra, where GPI-diagnostic ion matching as well as ω -site bearing C-terminal tryptic peptide identification are of great value for the final scoring. The purpose of VEMS-GPI predictor is slightly different comparing to other existing predictors as the experimentally obtained MS data will make the final decision.

All the analytical and bioinformatic methods presented in this chapter are generating a more solid foundation for the characterization of GPI-APs and for the understanding of their implication in cellular processes.

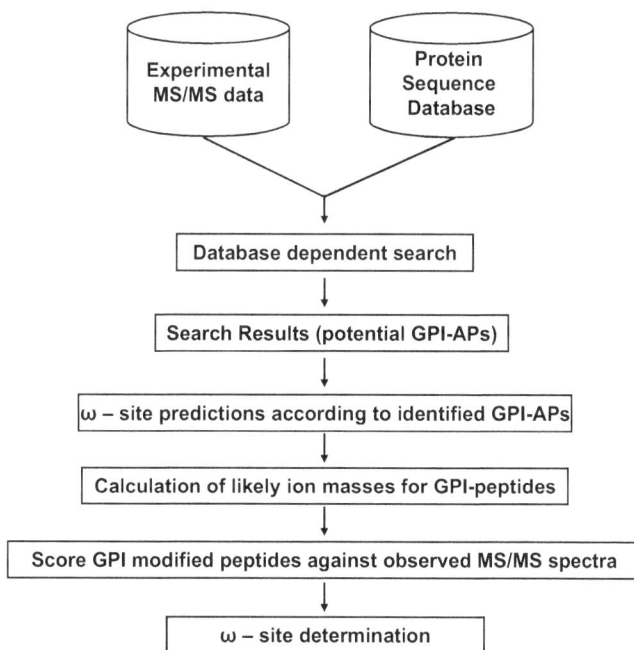

Figure 6: General scheme of the computational approach for identification and characterization of GPI-anchored peptides in proteomics experiments by VEMS-GPI.

REFERENCES

[1]   Jensen, O. N. Interpreting the protein language using proteomics. Nat Rev Mol Cell Biol 7, 391-403 (2006).

[2]   Eisenhaber, B. et al. Post-translational GPI lipid anchor modification of proteins in kingdoms of life: analysis of protein sequence data from complete genomes. Protein Eng 14, 17-25 (2001).

[3]   Orlean, P. & Menon, A. K. Thematic review series: lipid posttranslational modifications. GPI anchoring of protein in yeast and mammalian cells, or: how we learned to stop worrying and love glycophospholipids. J Lipid Res 48, 993-1011 (2007).

[4]   Hooper, N. M. Determination of glycosyl-phosphatidylinositol membrane protein anchorage. Proteomics 1, 748-55 (2001).

[5]   Eisenhaber, B. et al. Prediction of potential GPI-modification sites in proprotein sequences. J Mol Biol 292, 741-58 (1999).

[6]   Fankhauser, N. & Maser, P. Identification of GPI anchor attachment signals by a Kohonen self-organizing map. Bioinformatics 21, 1846-52 (2005).

[7]   Poisson, G. et al. FragAnchor: a large-scale predictor of glycosylphosphatidylinositol anchors in eukaryote protein sequences by qualitative scoring. Genomics Proteomics Bioinformatics 5, 121-30 (2007).

[8]   Pierleoni, A. et al. PredGPI: a GPI-anchor predictor. BMC Bioinformatics 9, 392 (2008).

[9]   Brown, D. A. & Rose, J. K. Sorting of GPI-anchored proteins to glycolipid-enriched membrane subdomains during transport to the apical cell surface. Cell 68, 533-44 (1992).

[10]   Dotti, C. G. et al. Polarized sorting of glypiated proteins in hippocampal neurons. Nature 349, 158-61 (1991).

[11]   Foster, L. J. et al. Unbiased quantitative proteomics of lipid rafts reveals high specificity for signaling factors. Proc Natl Acad Sci U.S.A. 100, 5813-8 (2003).

[12]   Hooper, N. M. et al. Renal dipeptidase is one of the membrane proteins released by phosphatidylinositol-specific phospholipase C. Biochem J 244, 465-9 (1987).

[13]   Deeg, M. A. & Davitz, M. A. Glycosylphosphatidylinositol-phospholipase D: a tool for glycosylphosphatidylinositol structural analysis. Methods Enzymol 250, 630-40 (1995).

[14]   Scallon, B. J. et al. Primary structure and functional activity of a phosphatidylinositol-glycan-specific phospholipase D. Science 252, 446-8 (1991).

[15]   Bordier, C. Phase separation of integral membrane proteins in Triton X-114 solution. J Biol Chem 256, 1604-7 (1981).

[16]   Borner, G. H., Lilley, K. S., Stevens, T. J. & Dupree, P. Identification of glycosylphosphatidylinositol-anchored proteins in Arabidopsis. A proteomic and genomic analysis. Plant Physiol 132, 568-77 (2003).

[17]   Elortza, F. et al. Proteomic analysis of glycosylphosphatidylinositol-anchored membrane proteins. Mol Cell Proteomics 2, 1261-70 (2003).

[18]   Sherrier, D. J. et al. Glycosylphosphatidylinositol-anchored cell-surface proteins from Arabidopsis. Electrophoresis 20, 2027-35 (1999).

[19]   Elortza, F. et al. Modification-specific proteomics of plasma membrane proteins: identification and characterization of glycosylphosphatidylinositol-anchored proteins released upon phospholipase D treatment. J Proteome Res 5, 935-43 (2006).

[20]   Aebersold, R. & Mann, M. Mass spectrometry-based proteomics. Nature 422, 198-207 (2003).

[21]   Steen, H. & Mann, M. The ABC's (and XYZ's) of peptide sequencing. Nat Rev Mol Cell Biol 5, 699-711 (2004).

[22]   Eisenhaber, B. et al. Sequence properties of GPI-anchored proteins near the omega-site: constraints for the polypeptide binding site of the putative transamidase. Protein Eng 11, 1155-61 (1998).

[23]   Eisenhaber, F. et al. Prediction of lipid posttranslational modifications and localization signals from protein sequences: big-Pi, NMT and PTS1. Nucleic Acids Res 31, 3631-4 (2003).

[24]   Udenfriend, S. & Kodukula, K. How glycosylphosphatidylinositol-anchored membrane proteins are made. Annu Rev Biochem 64, 563-91 (1995).

[25]   Ferguson, M. A. et al. Glycosyl-phosphatidylinositol moiety that anchors Trypanosoma brucei variant surface glycoprotein to the membrane. Science 239, 753-9 (1988).

[26]   Haas, R. et al. Glycoinositol phospholipid anchor and protein C-terminus of bovine erythrocyte acetylcholinesterase: analysis by mass spectrometry and by protein and DNA sequencing. Biochem J 314 (Pt 3), 817-25 (1996).

[27]   Roberts, W. L. et al. Structural characterization of the glycoinositol phospholipid membrane anchor of human erythrocyte acetylcholinesterase by fast atom bombardment mass spectrometry. J Biol Chem 263, 18776-84 (1988).

[28]   Fankhauser, C. et al. Structures of glycosylphosphatidylinositol membrane anchors from Saccharomyces cerevisiae. J Biol Chem 268, 26365-74 (1993).

[29]   Brewis, I. A. et al. Structures of the glycosylphosphatidylinositol anchors of porcine and human renal membrane dipeptidase. Comprehensive structural studies on the porcine anchor and interspecies comparison of the glycan core structures. J Biol Chem 270, 22946-56 (1995).

[30]   Hagglund, P. et al. A new strategy for identification of N-glycosylated proteins and unambiguous assignment of their glycosylation sites using HILIC enrichment and partial deglycosylation. J Proteome Res 3, 556-66 (2004).

[31]   Boersema, P. J. et al. Evaluation and optimization of ZIC-HILIC-RP as an alternative MudPIT strategy. J Proteome Res 6, 937-46 (2007).

[32]   Omaetxebarria, M. J. et al. Isolation and characterization of glycosylphosphatidylinositol-anchored peptides by hydrophilic interaction chromatography and MALDI tandem mass spectrometry. Anal Chem 78, 3335-41 (2006).

[33]   Larsen, M. R. et al. Highly selective enrichment of phosphorylated peptides from peptide mixtures using titanium dioxide microcolumns. Mol Cell Proteomics 4, 873-86 (2005).

[34]   Omaetxebarria, M. J. et al. Titanium dioxide, a promising approach for GPI-anchored peptide enrichment. HUPO 4th annual World Congress, 66 (2005).

[35]   Omaetxebarria, M. J. et al. Computational approach for identification and characterization of GPI-anchored peptides in proteomics experiments. Proteomics 7, 1951-60 (2007).

CHAPTER 5

Chemical Synthesis, Modification and Mimicry of the GPI Anchor

Martin J. Lear, Bastien Reux and Karthik Sekar

Abstract: Glycosylphosphatidylinositol (GPI) anchors are complex glycolipids, which typically anchor extracellular proteins onto the lipid membranes of eukaryotic cells. Although providing a natural platform in which to present or transfer functional molecules onto cells and viruses, GPI anchors are difficult biologics to generate in a homogenously pure form. It is also difficult, though not impossible, to elucidate and confirm their structures unambiguously. Today, chemical synthesis offers not only the versatility to make both complex and simplified GPI mimics and tools, but also the means to directly relate an exact GPI structure to its biological function. These synthetic GPIs may be further modified to allow the chemical attachment of any functional molecule, and not solely proteins, in a biologically compatible manner. Fluorescent labels and affinity tags can be exploited to investigate a particular biological response or process. Alternatively, synthetic glycans of GPI anchors can be employed to elicit a particular immune response or to generate GPI-specific antibodies. In this chapter, we shall overview the structure and synthesis of GPI anchors, and give perspectives on the biological study and therapeutic potential of synthetically-derived GPI biologics.

SYNTHESIS OF A FIELD

The chemical synthesis of the biologics of cells is an emerging field of study, one which has significant value in the understanding and treatment of a wide variety of diseases. Traditionally, organic chemists isolate, propose the molecular structures, and then verify through total synthesis, the secondary metabolites (the natural products) of plants, sponges, fungi, algae and bacteria. The chemistry behind making such natural products has changed significantly over recent years and has contributed greatly to our understanding of diseases and how medicines work [1,2]. But: what about the primary products or components of cells? What about making the biologics of cells; for example, the functional bioconjugates between nucleic acids, proteins, carbohydrates, and lipids? Can they be synthesised? If so, how can you make them in a pure form?

One primary obstacle in the synthesis of biologics stems from being able to know the exact structure to make. A second is the complexity. Multiple, yet closely related bioconjugated structures exist within cells [3]; for example, as a biomolecule departs from its genomic and proteomic origins, and experiences the post-translational glycomic, lipidomic, and metabolomic machinery, its complexity and diversity naturally increases, often dramatically. Within this dynamic cellular milieu, the unambiguous assignment of a molecular entity to a particular function (either singly or as a multivalent assembly) is an important goal to achieve. This correlation is crucial when studying the protein conjugates of the glycosylphosphatidylinositolsyl (GPI) anchor, whereby biologically derived samples (arguably heterogenous in chemical composition and connective integrity) are defined as "pure" in a functional context, but not in a structural one.

With regards to high resolution NMR, MS, and X-ray spectroscopy, there is arguably a lack of definitive structural characterization data within the glycobiology field that is traditionally a pre-requisite within the fields of natural product chemistry and carbohydrate chemistry. The disciplines of total and analogue chemical synthesis now promises to provide definitive links between the structures and functions of biologics, in particular, the glyco-conjugates of lipids and proteins. In discerning host-pathogen biology, for example, the structural detailing of complex glycophospholipids with a particular pathogenic response is required (Fig. **1**) [4,5]. Today, the total synthesis of exactly connected and stereochemically defined lipidated and glycan forms of the phosphatidylinositols mannosides (PIMs) provides pure homogeneic materials for biological study [6-9]. Amongst carbohydrate masters like Fraser-Reid [10-12], Boons [13], Kunz [14], and Schmidt [15-17], recent leaders in making synthetic glycolipids and glycoproteins include Danishefsky [17-19], Seeberger [20-23], Bertozzi [24,25], Davis [26,27], Wong [28,29], and Gin [30,31].

John A. Dangerfield & Christoph Metzner (Eds)

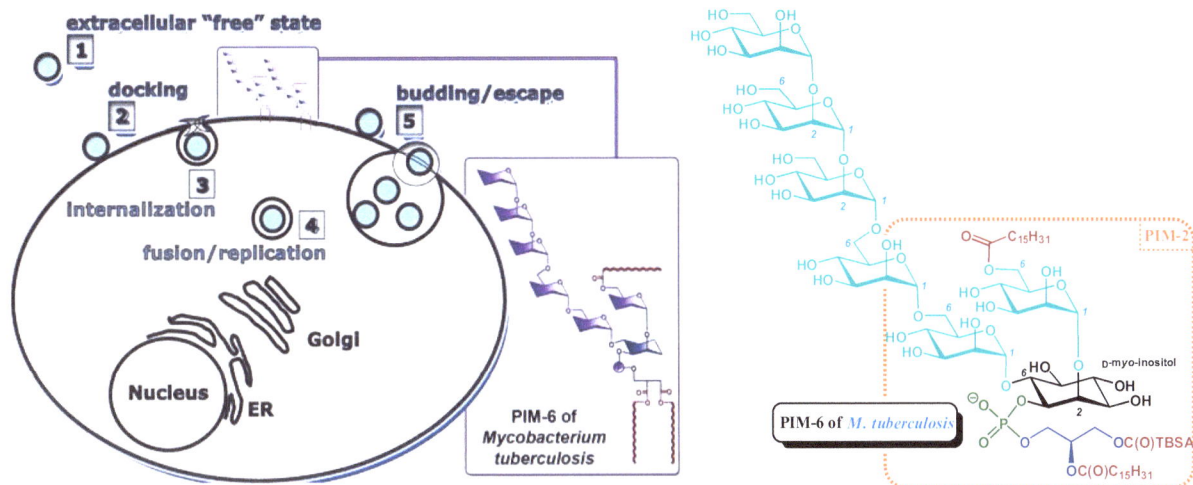

Figure 1: Glycolipids of a pathogenic organism. As represented by membrane-anchored phosphatidylinositol mannosides of *Mycobacterium tuberculosis* (TBSA= ester of tuberculostearic acid, i.e., (10*R*)-methyloctadecanoic acid).

The chemical synthesis of glycolipid analogues offers the further promise to pin-point recog-nition and trafficking events of intracellular lipidated pathogenic molecules of *Mycobacterium tuberculosis* and *Plasmodium falicparum*. This can include, for example, the development of biomarker-standards to analyze infection, antigens to provide glycolipid antibodies, and assay baits to identify enzymes. These are a few of the possibilities to making synthetic glycolipids. The power of chemical synthesis is now ready to take on and add molecular clarity to the biologics field (Fig. **1**). In this chapter, we shall focus on the structural issues and synthetic potential surrounding the membrane-bound glycolipids known as the glycosylphosphatidylinositol (GPI) anchors.

CHEMICAL STRUCTURES OF GPIS

Although postulated by Golgi in 1896, the first GPI structure was characterized by Ferguson *et al.* in 1988 from the parasite *Trypanosoma brucei*, which can remarkably survive host blood-streams because it is densely coated with GPI-anchored proteins, named variant surface glyco-proteins (VSG) [32]. Even when stripped down, the GPI-anchor is truly complex, being composed of several domains: a phosphoethanolamine (PEtN) for protein-attachment, a variably glycosylated poly-mannoside (glycan) chain off a D-glucosamine (GluN) residue and a phosphatidylinositol (PI) lipid (diacyl glycerol) base for membrane-binding (Fig. **2**) [33,34]. From species to species, the GPI-anchor can differ significantly in structure, and can bear additional or yet more diversified glycan and lipid branches at various positions (Fig. **3**). Typically, however, the GPI anchor is attached to the C-terminal end of a protein [35] or a polysaccharide like the phosphoglycans of Leishmania parasites [36-38]. GPI anchoring amongst these latter protozoans shows interesting evolutionary connections in their species-, stage-, and tissue-specific GPI structures [39,40]. Higher eukaryotes make use of GPI anchoring principally to present functionally diverse proteins that play key roles in signal transduction, immuno-pathogenesis and cell-to-cell contact behaviour, for example, within the blood and connective tissues [41-46]. Every major cell type and tissue in vertebrates contains a GPI-anchored protein. To date, more than 250 proteins have been identified. Each can vary in size, ranging from 12 amino acids for the cluster of differentiation (CD) antigen, CD52 peptide, to the 175 kDa protein of CDw109. Defects in GPI anchor biosynthesis can result in embryonic lethality in mammals and yeast [47,48].

Despite being isolated in a wide variety of mammalian tissues, for example, the squid brain, the slime mold *Dictyostelium*, the fungal yeast *Saccharomyces*, and in many protozoan parasites, opinion is divided as to the presence of GPIs in prokaryotes, bacteria or algae [39-42,49-65]. The structural diversity of eukaryotic GPI anchors is mainly reflected in the location and nature of the branching groups off the poly-mannoside (glycan) chain. As aptly commented by Peter H. Seeberger, this is where the structural language between the nomenclature of glycobiologists and the molecular diagrams of chemists can become confusing or perhaps generate ambiguity [20]; for example, the GPI anchor of *Trypanosoma cruzi* may be described as Manα1-2(ethanolamine phosphate)Manα1-2Manα1-6Manα1-4(2-aminoethylphosphonate-6)GlcNα1-6-*myo*-inositol-1-*P*-3(*sn*-1-*O*-(C16:0) alkyl-2-*O*-(C16:0) acylglycerol) [65]. Both methods of structural description are given here for comparative reasons (Fig. **3**). The need for uniformity and structural exactitude within the glycolipid and glycopeptide fields is certainly clear.

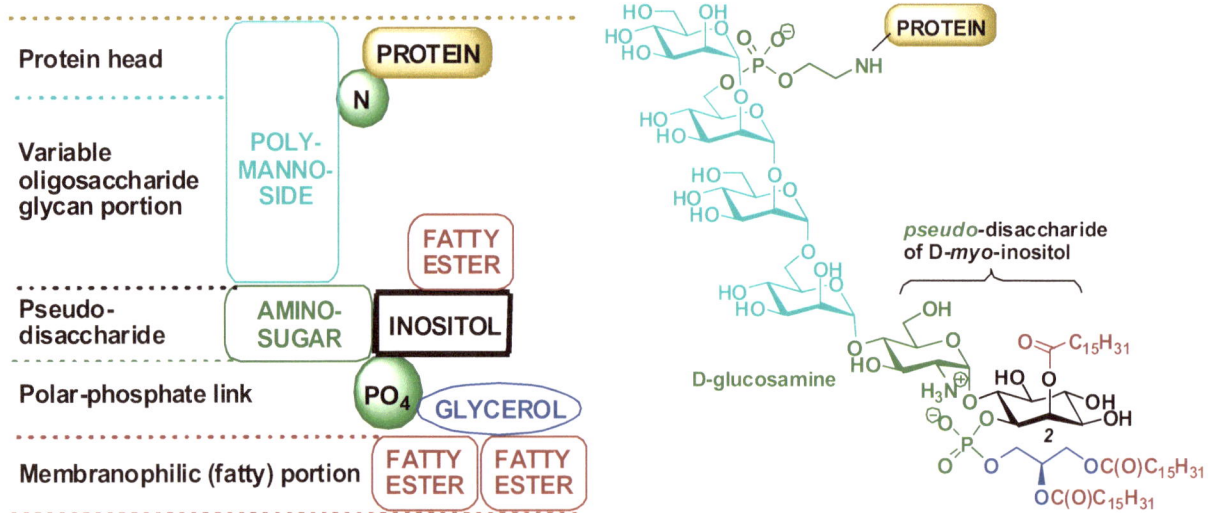

Figure 2: The structural domains of a protein-bound GPI anchor of *P. falciparum*.

Referring to Fig. **3**, higher eukaryotes can bear additional phosphoethanolamine (PEtN) residues (*e.g.*, the GPI anchor of rat brain Thy-1) [43]. While many GPI anchors contain a diacylglycerol lipid moiety (*e.g.*, the *sn*-1,2-dimyristoylglycerol in *Trypanosoma brucei* VSG) [32], certain alkylacylglycerol residues are also found in, for example, human AchE [49] and rat brain Thy-143. Ceramide-based lipids have also been identified in *Saccharomyces cerevisiae* [56]. The human CD52 antigen is a GPI anchored glycopeptide antigen involved in both human reproduction and immune recognition processes [59,60]. The most variable region within GPI anchors typically reside at the C4 and/or C3 position of the mannose that is directly attached to the D-glucosamine. The glycan A and B antigens of *Toxoplasma gondii*, for example, are part of a family of protein-free GPI glycolipids, whereby glycan A features a GalNAc linked α(1-4) to the mannose and glycan B features a Glcβ(1-4)GalNAcα(1-4) side-branch [57].

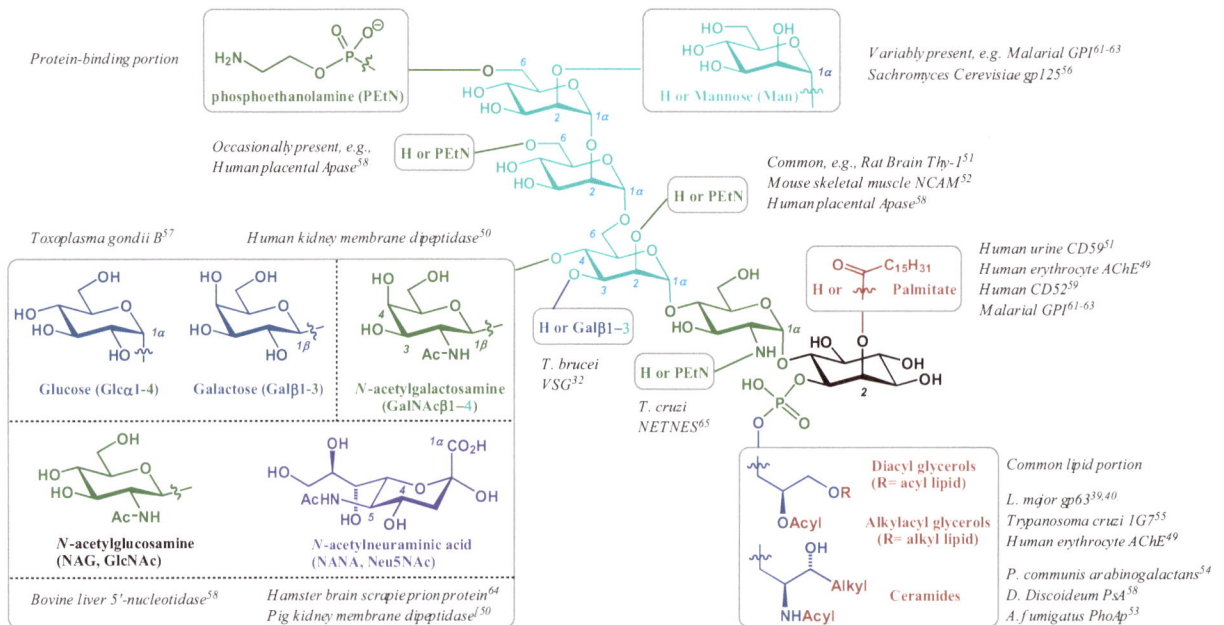

Figure 3: Molecular overview of glycan and lipid heterogeneity within selected GPI anchors. Each box represents a region of known diversity. The referenced GPIs typically bear multiple regions of structural modification and variability. The variable glyceride or ceramide lipids, and the C4 or C3 sugars off the third mannose unit, are not always known with certainty; for example, the lipids of the GPI anchor of the Hamster brain scrapie prion protein remain to be determined [64], and GalNAc☐1-4 oligosaccharide side-branches vary in length and connectivity with NANA, Gal, and/or GalNAcβ1-4 sugars.

TOTAL SYNTHESIS OF GPI ANCHORS

Total synthesis is a field of organic chemistry that involves the construction of Nature's diverse and complex assortment of molecular structures from simple, commercial building blocks. Organic chemists in this field first of all need to know the structural puzzle to build. Blueprints can then be designed and solved on the bench, and the resultant molecule finally realised. Occasionally, a structural re-assignment of the natural product is necessary after the completion of a total synthesis [66,67]. For glycolipids, the sugar and lipid pieces of the puzzle are first made and strategically functionalised, so that they can be connected (reacted) in structurally and stereochemically defined ways (Fig. **4**). Benzyl (Bn) ether protecting groups, which block sugar positions from being lipidated, phosphorylated or glycosylated during the assembly process, are typically removed (deprotected) at the last step of the synthesis.

Several total syntheses of GPI anchors have been reported in the chemistry literature. Fig. **4** summarises a common unified strategy, whereby the GPI glycan is first constructed in an [n+2] fashion between the variable (n) oligosaccharide chain and the glucosamine-inositol (pseudodisaccharide) unit, after which the (n+2)-glycan is coupled (phospholipidated) to the phospholipid glycerol or ceramide base. The group of Ley, for example, have developed a quick and versatile (5+2)-glycan assembly of the GPI anchor of *Trypanosoma brucei,* an African parasite that causes Sleeping-sickness in humans and domestic animals [68,69]. The pseudo-disaccharide unit was first constructed from a D-glucosamine azido-equivalent and a D-*myo*-inositol that was desymmetrised in an enantio-enriched form (*ee* ≥ 98%) as a dispiroketal [70-72]. The trisaccharide was 6-*O*-glycosidated with a disaccharide selenophenyl donor under methyl triflate (MeOTf) conditions to produce the pentasaccharide chain in 75% yield. Activation of the pentasaccharide thioethyl donor with *N*-iodosuccinimide and trifluoromethane sulfonic acid (NIS/TfOH) led to the glycosidation of the inositol pseudodisaccharide fragment and furnished the heptasaccharide (5+2)-glycan core in 51% yield. The synthesis was completed by attaching the phosphoramidite diacyl lipid portion and removing all protecting groups to afford the GPI anchor. Various other groups have also achieved the total [73-75] and partial [76-78] synthesis of the GPI anchor of *T. brucei.*

To discern the role of GPIs in malaria pathology and signal transduction, and to investigate GPI biosynthesis, access to the pure, naturally occurring, lipidated malarial GPI form is of paramount importance. Using chemical synthesis, the group of Seeberger [79] have identified the glycan part of the malaria toxin, a cell surface GPI from *P. falciparum* that is largely responsible for an alarmingly high malaria mortality rate [80]. The group further demonstrated that a synthetic glycan can serve as an effective antitoxin vaccine in a rodent model [81]. Their total synthesis of a malarial GPI candidate centred around the (4+2)-glycan coupling of a suitably-protected tetramannoside chain [72] and a glucosamine-inositol pseudodisaccharide (Fig. **4**). The hexasaccharide glycan core was then sequentially and specifically decorated with lipids, phospholipids and phosphoethanolamine (PEtN) followed by global deprotection to afford the malarial GPI anchor. An alternative total synthesis approach includes that of Fraser-Reid [73]; other partial [74-87] and semi-syntheses [88] have also been achieved.

The first mammalian GPI to be isolated, the GPI anchor of rat brain Thy-1, was first synthesized by the group of Schmidt in 2003 (Fig. **4**) [89]. The pseudoheptasaccharide (5+2)-glycan chain was assembled by trichloroacetimidate coupling of the pre-assembled (3+2)-pentasaccharide portion to the tert-butyloxycarbonyl (Boc) nitrogen-protected glucosamine-inositol pseudodisaccharide in 74% yield by using trimethylsilyl triflate (TMSOTf) as the coupling reagent. The heptasaccharide was then decorated with the PEtN moiety and phospholipid diester base to afford the GPI anchor of the rat brain Thy-190. Notably, the group of Fraser-Reid have developed a shorter total synthesis of the rat brain Thy-1 GPI anchor using their clever n-pentenyl orthoester methodology [91-93].

CD52 is a GPI-anchored glycopeptide antigen associated with the recognition processes of the human immune and reproductive systems. Monoclonal antibodies of the human lymphocyte CD52 and sperm CD52, for example, have shown promise in the development of immune-related [94-100] and contraceptive [101,102] treatments, respectively. Wu and Guo [103] have devised a (3+2)-glycan strategy to assemble the GPI anchor of the human CD52 antigen of sperm by the α-stereoselective anomeric coupling of a pre-formed trimannose trichloroacetimidate unit [104] to the 4-*O*-position of a glucosamine pseudodisaccharide [105,106] of *myo*-inositol bearing a palmitoyl group (Fig. **4**). Their orthogonal deprotection strategy then allowed for the selective attachment of the PEtN and the phospholipid moieties before global benzyl deprotection uncloaked the full GPI anchor [103]. A few partial and semi-syntheses have also been performed by the same group [107-109]. Other syntheses of GPI anchors include those from the groups of Konradsson (GPI anchor of *T. cruzi*) [110-112], Schmidt (GPI anchor of *Sacchromyces cerevisia*) [113], and Ma (two GPI anchor glycans of *S. cerevisiae* and *A. fumigates*) [114].

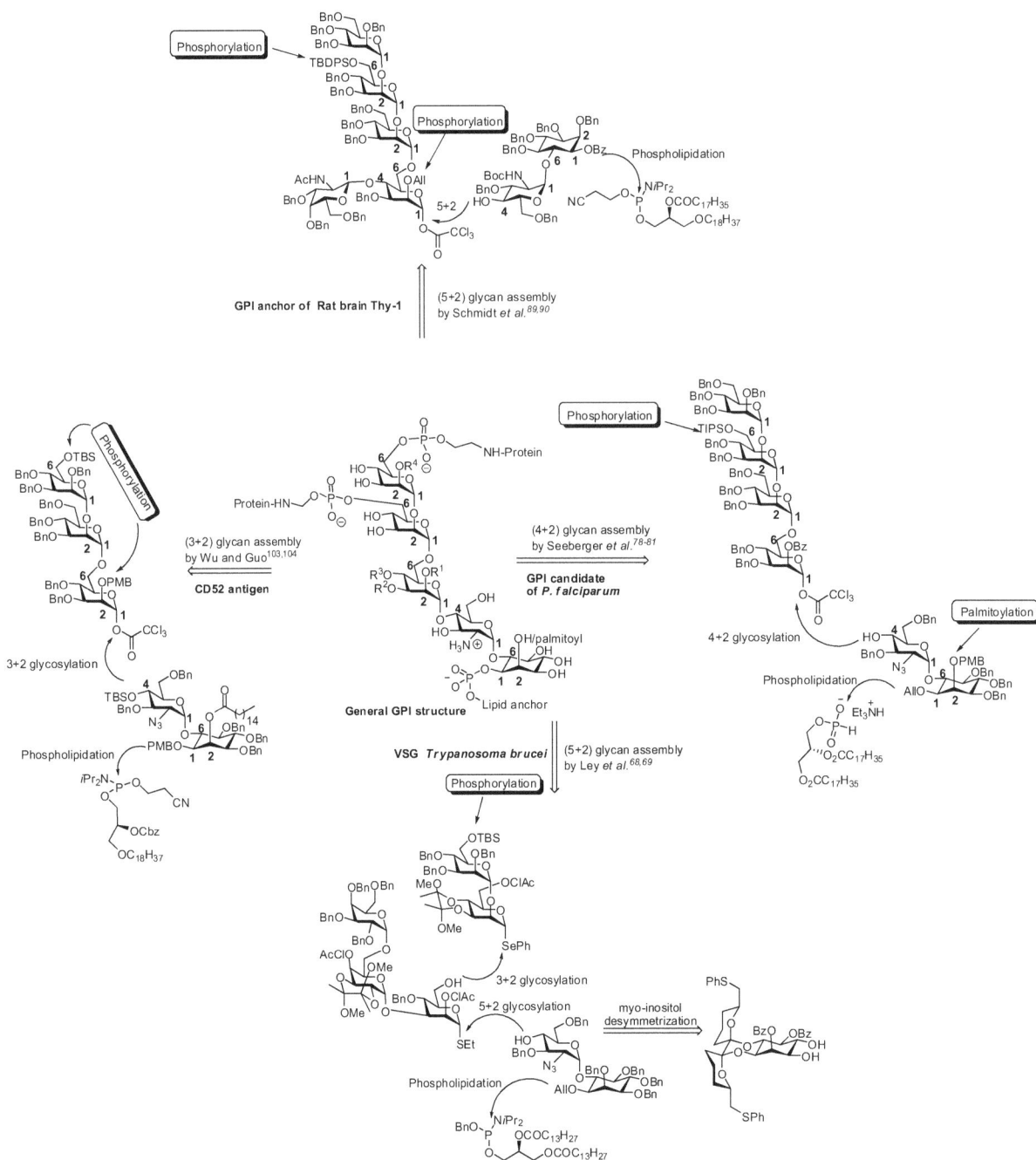

Figure 4: A unified total synthesis blueprint: the synthetic assembly of selected pieces of several GPI puzzles, which have been successfully prepared within the chemistry community.

SYNTHETIC GPI ANALOGUES

Within the signal transduction processes of cells, it is generally accepted that GPI anchors target lipid rafts, specific intracellular compartments, and the apical membranes of polarized epithelial cells [115]. But: which of the diversified range of GPI protein conjugates does what? And: how can we investigate such biological phenomena precisely? Access to specific and homogenous GPI structures would certainly help. Yet, without knowing all the biosynthetic enzymes and post-translational machinery [116-121], it is arguably beyond the reach of contemporary biology to prepare a specific GPI structure in a confident manner. The systematic and general principles of chemical synthesis can now provide the needed techniques and tools to make truly specific GPI structures; not only this, but also the capability to discern such biosynthetic and post-translational processes by providing precise GPI structural variants and metabolic GPI precursors for study in this emerging field.

As highlighted by Paulick and Bertozzi, a growing number of groups have already begun to study and mimic an extensive range of natural GPI anchors using chemical synthesis [122]. Typically, simplified GPI or glycolipid variants are made and attached (or lipidated) to synthetic, recombinant or expressed peptides and proteins. The synthetic methods involved are varyingly termed protein lipidation [123], peptide ligation [124] or native chemical ligation (NCL) [125,126]. To organic chemists and molecular biologists alike, the 'site-specific' lipidation of proteins [127] and the C-terminal grafting of proteins onto synthetic GPIs are still formidable challenges [128]. The group of Moroder has studied the C-terminal lipidation of proteins by employing 'click-chemistry' [129,131], the Huisgen 1,3-dipolar cycloaddition of azides to terminal alkynes to form stable triazole products (Fig. **5**) [132]. Later the product was transthioesterified between *N*-cysteinyl lipopeptides (fluorescently labeled with *N*-rhodamine B and *N*-dansyl) and peptide thioesters. Incubation of HeLa cells with the lipopeptide attached to the Rhodamine B revealed a fast uptake and intracellular membrane distribution within 30 minutes, and eventual cytoplasmatic (proteolytic) release of the fluorescent dye [133,134].

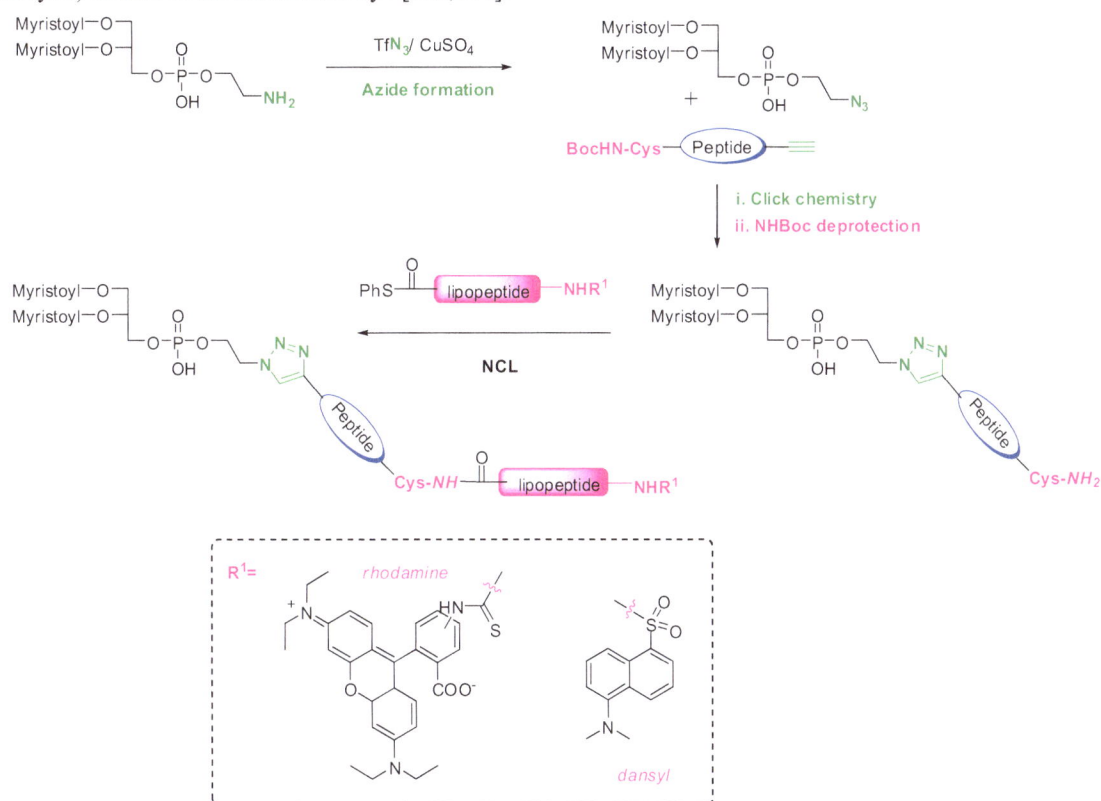

Figure 5: Use of native chemical ligation (NCL). NCL is a technique coined 'click chemistry' [129-131] for N-terminally fluorescent-labelled and C-terminally lipidated model peptides [132-134].

In contrast to cytosolic amyloidogenic proteins associated with neurological disorders, the GPI anchor can modulate the polymerization of prion proteins (PrP) and the concentration of amyloid fibrils through self-assembly processes on cellular membranes [135,136]. By mimicking the prion GPI anchor with a cysteine-conjugated myristoyl-maleimide attached to the C-terminal residue of recombinant-PrP (rPrP), the Baskakov group demonstrated that a GPI mimic (myr-PrP-(S230C)) can pack polypeptide prion chains into a cross β-sheet fibril form, without which an infectious state would not occur (Fig. **6A**) [137-139]. GPI-deficient PrP polypeptides may therefore adopt different conformations and assemblies from disease-associated forms. Similarly, in the fluorescent-guided investigation of lipidated prion proteins in human epithelial kidney and mouse neuronal cells, the Becker group tailor made different rPrP peptides bearing N-terminal cysteine residues for ligation to maleimide GPI-mimics (Fig. **6B**) [140-142]. The peptides were designed and palmitated with and without TEV, the Tobacco etch virus amino acid recognition site ENLYFQ.

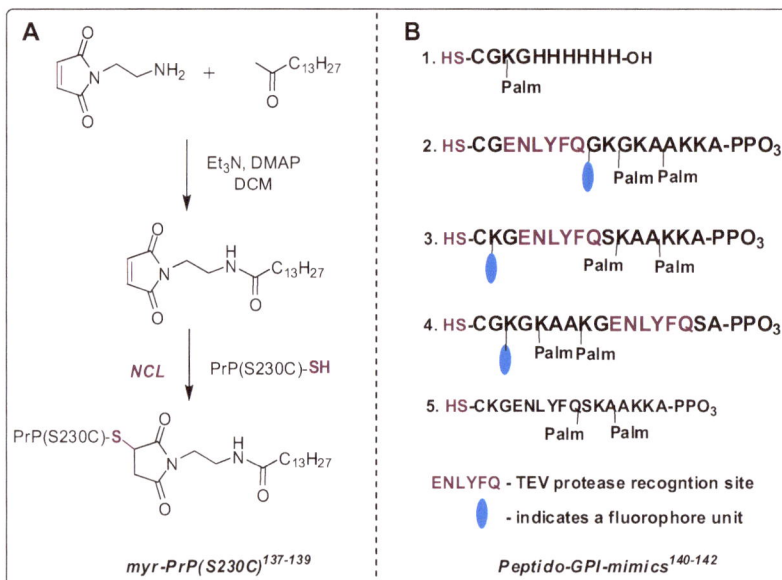

Figure 6: Synthesis of GPI-like conjugates via native chemical ligation (NCL) between cysteine-terminated peptides [137-139] and myristoyl-linked maleimides [140-142].

In a significant extension of their earlier work on lipidated peptides as GPI-mimics [143], the Bertozzi group have synthesised a range of GPI analogues by mimicking the glycan core of the GPI anchor with mannosyl terminated polyethylene glycol (PEG) chains of appropriate lengths

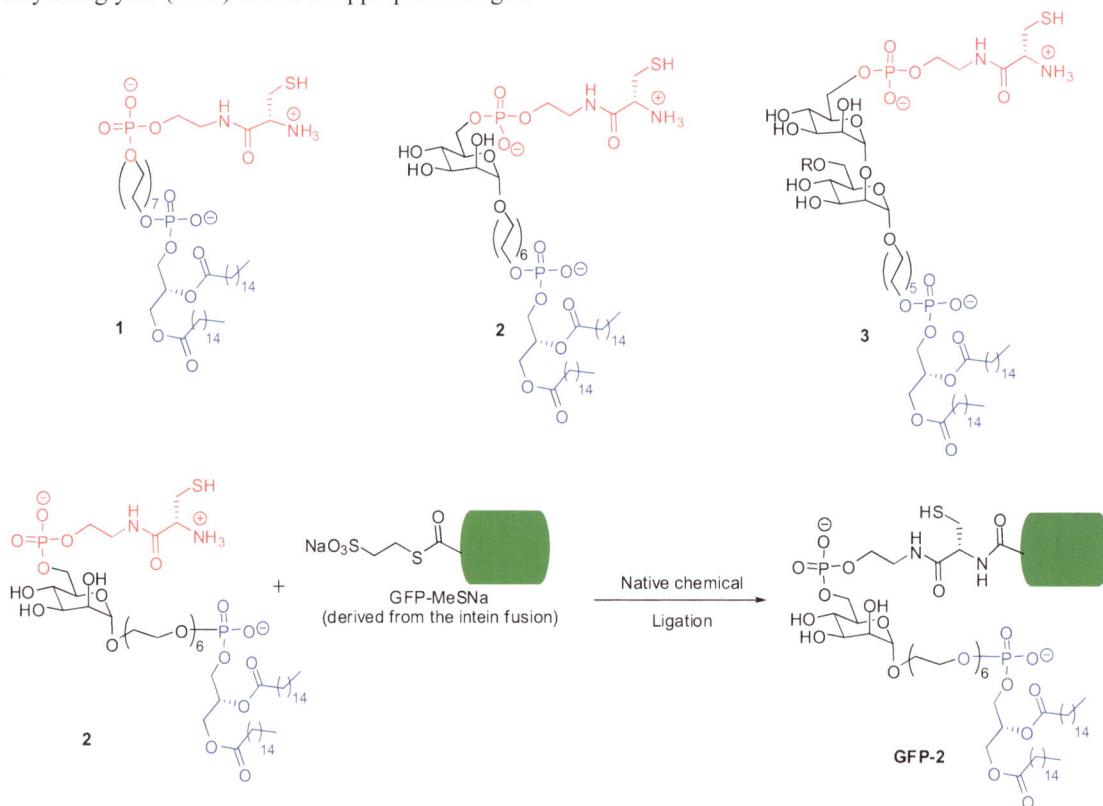

Figure 7: Representative GPI anchor analogues from the Bertozzi group, comprising three domains: PEtN linker (red), glycan core (black), phospholipid tail (blue) [144]. Adapted under copyright ©2007 National Academy of Sciences, U.S.A.

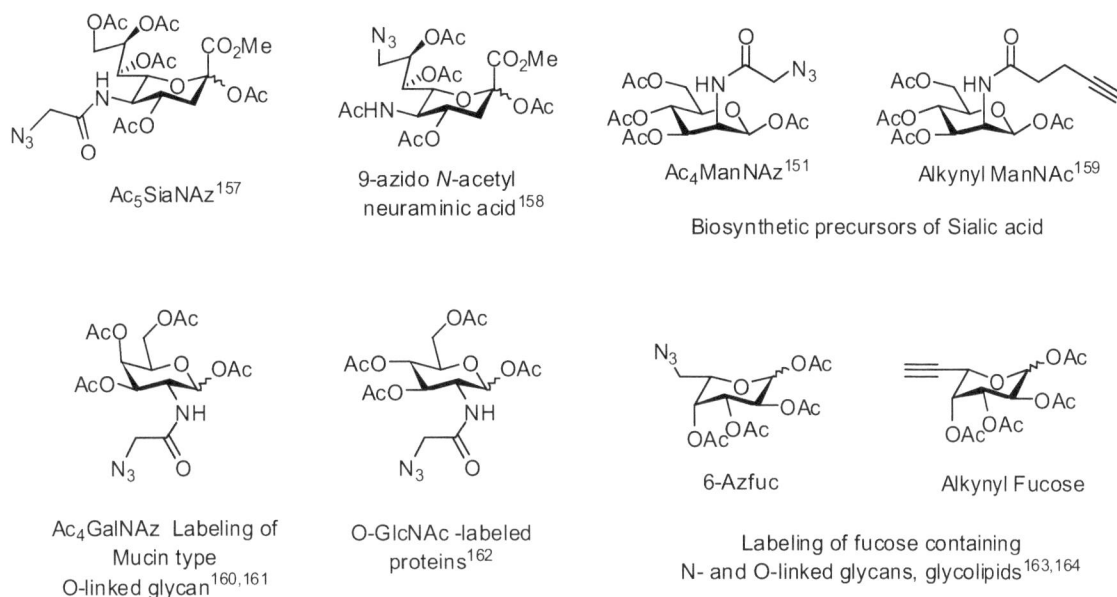

Figure 8: Synthetically modified sugars with azide and alkyne functionality that have found some success in metabolic incorporation and labelling studies [150]. Adapted under copyright ©2009 National Academy of Sciences, U.S.A.

(Fig. **7**) [144]. The modified PEtN-cysteine moiety allowed ligation (NCL) to the thioester-activated, recombinant green fluorescent protein (GFP) [145]. Similar to other GPI-protein congeners [146-149], these exogenous GFP-tagged GPI-mimics incorporate into the mammalian cell membrane of living cells, as analysed by flow cytometry and optical imaging techniques. Although membrane mobility was affected by the monosaccharides of the glycan core, minimal perturbation to the cells and endocytosis of the fluorescent GPI-protein (GFP) have been observed. GPI-mimics lacking a lipid portion did not incorporate into cell membranes. Those lacking sugars, but bearing lipid portions, did incorporate into mammalian HeLa cell membranes. Further studies, however, showed these GPI-mimics to possess lower lateral mobility, as compared to both synthetic GPI-mimics bearing sugars and native GPIs that have been endogenously labelled with GFP.

SYNTHETIC GPI BIOLOGY

There are several chemical methods to study glycolipids both outwith (*ex vivo*) and within (*in vivo*) living cells. These methods are often termed bioorthogonal, since the chemistry involved will conveniently react with a particular, synthetically-incorporated functionality rather than those found naturally. In their excellent review, Laughlin and Bertozzi [150] describe how glycans can be metabolically incorporated with synthetically-modified sugars (reporters) that bear bioorthogonal functionality (*e.g.*, azides or terminal alkynes). This process is sometimes called metabolic engineering (ME). Typically, a sugar (or its biosynthetic precursor) is modified and then offered to cells dependent on that particular sugar; after which, an NCL method (*e.g.*, Staudinger ligation [151] or 'click chemistry' [152-156]) is adopted to tag on a fluorescent or affinity label. Some unnatural monosaccharides, which have been successfully engineered into glycans are referenced in Fig. **8**. These analogues are per-*O*-acetylated to increase permeability into the mammalian cell membrane. The fucosyl derivatives have limited use, however, because of cytotoxicity and low levels of metabolic incorporation. To date, both the glycosaminoglycans and GPI anchors have failed to be metabolically modified with azido and alkynyl monosaccharides.

The metabolically cultured azide and alkyne-labeled glycans can be imaged optically by chemical reaction with fluorophore-conjugated phosphines or alkynes (Fig. **9A**). The cell surface SiaNAz residues, for example, have been visualised in live cells by the Staudinger ligation with rhodamine, fluorescein, and Cy5.5 conjugated phosphine reagents. The coumarin scaffold [165] has also been used as a phosphine probe with the limitation being the generation of phosphine oxide, which will, in turn, generate background fluorescence because of the nonspecific oxidation during Staudinger ligation. This has been overcome by adopting a red-shifted fluorogenic phosphine reagent [166]. Later the Wong group [159,164] introduced the 1,8-napthalimide scaffold (Fig. **9B**) to probe the fucosyl and mannosyl labeled glycans using copper (Cu) mediated click chemistry. The use of metal-free click chemistry can resolve such issues in live cell imaging studies.

Although azides and alkynes are able to react with other bioorthogonal functional groups under physiological conditions, the Cu(I)-catalyzed 'click chemistry' is cytotoxic and cannot be used to image fixed cells or tissues. To address this issue, the Bertozzi group developed a *gem*-difluorocyclooctyne (DIFO) reagent that can 'click' without the need for any metal catalyst (due to special electronic factors and the release of ring-strain) [154,155]. The glycans metabolically labelled with azide groups can thus be reacted with DIFO in a direct, immediate, and simple manner, with the result that cell trafficking can be monitored by fluorescence microscopy. This more gentle technique has even been applied to labelling and tracking glycans in live organisms. In zebra fish embryos, for example, galactosamine and mannosylamine azido derivatives can first be metabolically incorporated and then directly imaged with DIFO (Fig. **10**) [161].

SYNTHETIC GLYCAN ARRAYS

The preparation of immobilised arrays of immunogenic glycan ligands seems to be a recent trend in the literature, perhaps due to the growing popularity of automated oligosaccharide synthesis [22,167,168]. The chosen glycans can be derived from lipids or peptides, which are selected based on the disease or pathogen under study, for example, PIM-based glycans can be chosen for the study of malaria [169,170], GPI-based glycans for the study of *Mycobacterium tuberculosis* [171] or glycoprotein-based glycans for the study of HIV-1 [172]. Often it is the interaction of glycolipid proteins on the outer envelope of the pathogen with receptors of the mammalian host cell that is central to the early stages of infection. Notably, the Seeberger group has synthesised several PIM7 and GPI [173] arrays (Fig. **11**). This has revealed pertinent host cell interactions and immunogenic responses, for example, synthetic PIM-6 glycans only displayed an increase of T-cell proliferation and IFN-γ production, whereas the lower-order homologues interestingly did not.

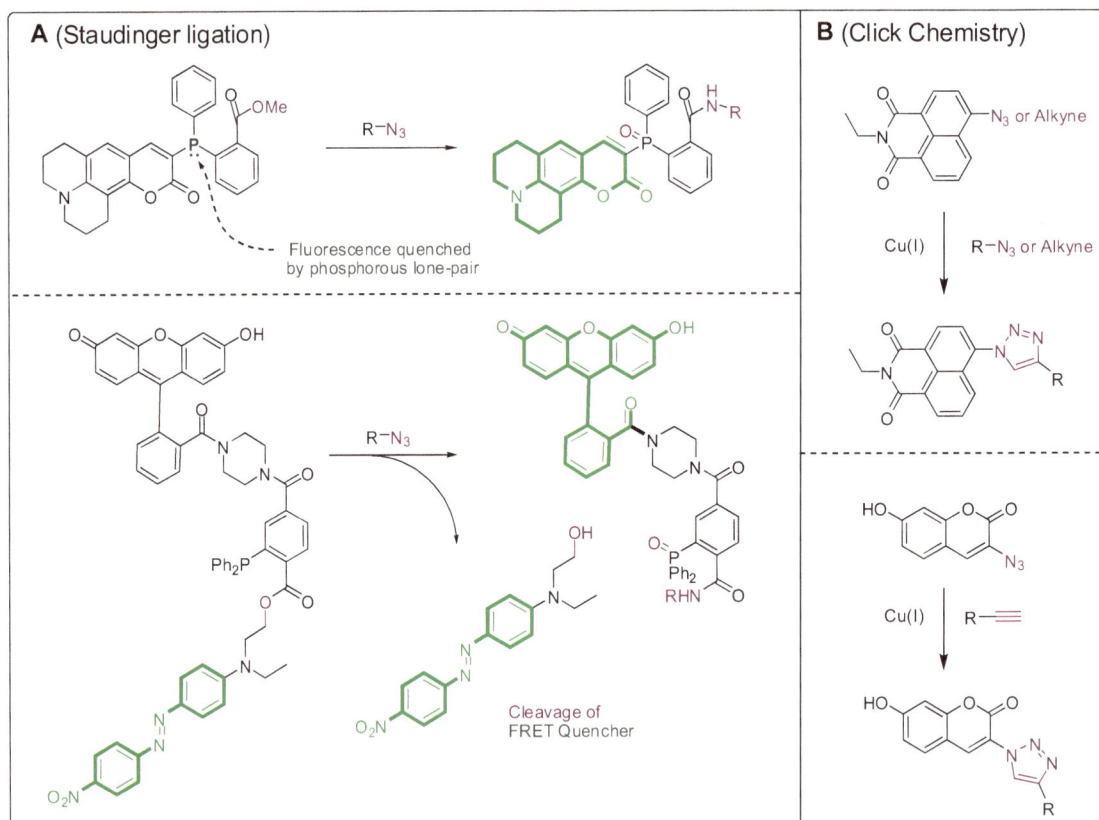

Figure 9: Fluorophore activation and labelling via Staudinger ligation (A) or click chemistry (B) [150]. Adapted under copyright ©2009 National Academy of Sciences, U.S.A.

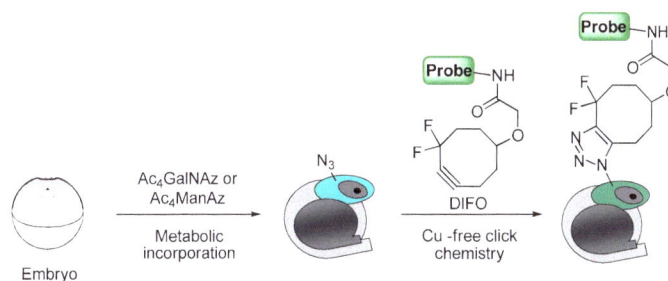

Figure 10: Metabolic engineering and imaging of Zebra fish embryos. Undertaken with Ac$_4$GalNAz followed by metal-free click chemistry with AlexaFluor647-conjugated DIFO and AlexaFluor488-conjugated DIFO [150]. Adapted under copyright ©2009 National Academy of Sciences, U.S.A.

In particular, the Seeberger group has established a minimal structure within their GPI-glycan arrays that is capable of distinguishing anti-malarial antibodies in serum samples (Fig. **11**). Their results indicate that glycans with longer mannose chains (GPI-6, GPI-7, GPI-8) show substantial binding ability to serum samples, whereas lesser carbohydrate units (GPI-1 to GPI-5) do not elicit detectable responses. The PetN group in GPI-5 has no significant binding ability. The key recognition element needed to generate a reactive GPI antibody response was found to be the fourth, capping mannose unit8. When higher-order GPI oligosaccharides (GPI-7 and GPI-8) were coupled to carrier proteins, strong immunogenic and protective responses against malarial pathogenesis and mortality were observed in a preclinical rodent model [9]. In related studies, the sialic acid-binding immunoglobulin-like lectin 8 (siglec-8) was identified [174] and carbohydrate microarrays for the human immunodeficiency virus type 1 (HIV-1) envelope proteins (gp120 and gp41) have been reported (Fig. **12**) [172].

Figure 11: Synthetic GPI-glycan microarray from the Seeberger group [8,9,23,173]. The glycan homologues were immobilised on a maleimide coated glass slide through NCL to the thiol linkages of varying lengths to mimic lipid anchoring. Furthermore, photo-affinity labels and carrier protein conjugates were similarly constructed via the NCL technique. Adapted by permission from Macmillan Publishers Ltd: Nature Chemical Biology [8], copyright ©2008.

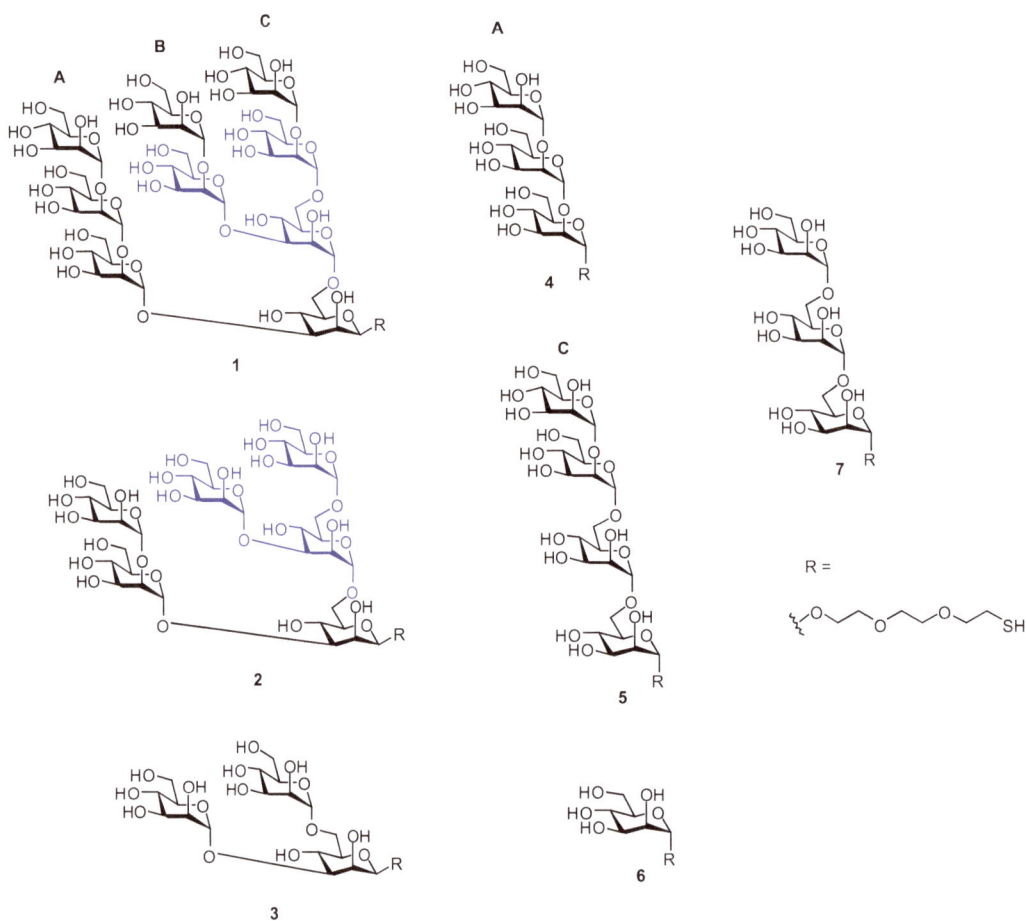

Figure 12: Oligosaccharide and glycoprotein microarrays produced to study the HIV-1 glycobiology of high-mannose type of oligosaccharides. Branched mannosides are coloured violet. Reproduced from Chemistry & Biology [172] with permission from Elsevier Ltd, copyright ©2004.

Carbohydrate arrays are not new to glycobiology and have been previously utilised in various biological and diagnostic ways, for example, to inhibit fucosyltransferases [175], galactosyl transferases [176] or sialyl transferases [177], or to study the binding affinity of fibroblast growth factor (FGF) to heparins [178] or glycosaminoglycans [179]. Glycan arrays of chondroitin sulfate oligosaccharides have been used to study the binding of brain-derived neutrotrophic factor (BDNF), growth factor midkine, FGF-1 and TNF-α[180,181]. Diagnostically, glycan arrays have been used to assess antibody levels in human sera, the binding preferences of haemogglutinins, and the presence of human H3N2 and avian H5N1 hemagglutinins [182]. Notably, instead of using the artificial support of glass slides, which are typically employed for making glycan arrays, the Harland group [183] has developed a synthetic lipid membrane to support trehalose glycolipids. Biomimetic technologies such as this have promise in the future study of glycolipid-based interactions.

PROSPECTS

In closing, we hope to have shown that the synthetic development of biologically relevant molecules is an important focus of chemistry [1,2]. In the later sections, various synthetic mimics or simplified portions of the GPI anchor were introduced. These synthetic GPI anchors can be used as cellular tools and biological probes to unravel or image the intricate processes and interactions of cells [10-27].

It is true that the generation of synthetically derived carbohydrate-based biologics is a growing field, especially regarding vaccines [19-21]. Evidence suggests, however, that it may be impractical to produce sufficient quantities of biological GPIs, and this is where synthetic chemical analogues of bioactive glycolipids of the GPI anchor will become invaluable for downstream medical applications. So, in the future: why not make synthetic GPI-mimics to carry therapeutic packages that can be presented on the outer membranes of cells or viruses? Can we exploit the GPI anchor as a platform for the targeted delivery of therapeutic drugs, functional proteins, or beneficial genes? These are exciting new therapeutic strategies to explore and develop in the future, and many aspects of this are discussed in detail in other chapters of this book.

Currently, one way to treat solid tumours harnesses special viruses that can invoke a therapeutic "killing" effect within cancer cells [184]. Unfortunately, however, some barriers still exist for using such vectors in the clinic, such as the evasion of the patient's immune system and the mistargeting and infection of unintended cells. Recent evidence strongly suggests such challenges may be overcome by modifying the surface of the virion with GPI-anchored proteins [185,186].

Could the coating of viruses with GPI anchored antibodies specific to the antigens of the targeted tissue help towards developing such medicines? How about painting magnetic nanoparticles (MNPs) onto the membranes of cells or viruses using GPI anchors? Why not design synthetic GPIs to allow tracking of cells or viruses within the body using imaging modalities such as magnetic resonance imaging (MRI), positron emission tomography (PET) or single photon emission computed tomography (SPECT)?

To date, these directions still remain to be solidified within the scientific community [187]. This will require the merging of chemical synthesis with GPI biology. Nevertheless, the *in vitro* manipulation of viruses with biologically-derived, GPI anchored MNPs (cf. Chapter 6) [185,186] and the *in vitro* imaging of synthetically-derived, fluorescent-labelled GPI-mimics (Fig. **13**) [188] has already begun.

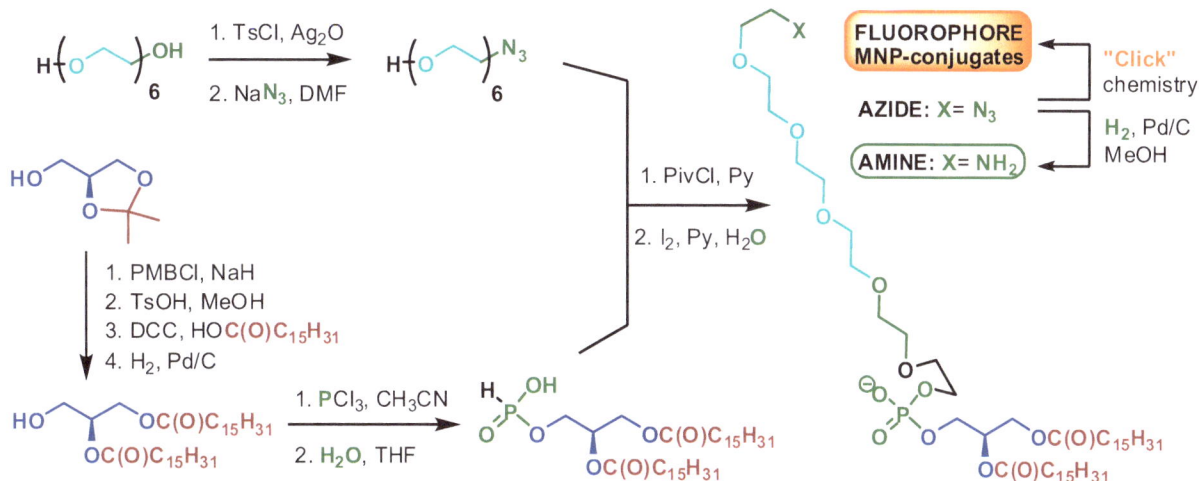

Figure 13: Synthetic route to a fluorescent-capped GPI-mimic for the exogenous painting of viruses and cells [188].

ACKNOWLEDGEMENTS

We thank the National Research Foundation (NRF) of Singapore for generous financial support under a Competitive Research Program (NRF-G-CRP 2007-04) that has allowed the chemistry and biology communities to find a common ground in the study of glycolipids. We further would like to dedicate this chapter to the 80th Anniversary of Chemistry and Science in Singapore (1929-2009).

REFERENCES

[1]   Montagnon, T. & Nicolaou, K. C. Molecules that Changed the World, Wiley-Blackwell; Singapore, March 2008.
[2]   Corey, E. J. et al. Molecules and Medicine, Wiley; New Jersey, U.S.A., August 2007.

[3]   Varki, A. et al. Essentials of Glycobiology, 2nd edition (Cold Spring Harbor Press Press, Cold Spring Harbor, New York, U.S.A., 2009).

[4]   Wenk, M. R. The emerging field of lipidomics, Nature Reviews: Drug Discovery 4, 594 (2005).

[5]   Wenk, M. R. Lipidomics of host-pathogen interactions, FEBS Lett 580, 5541(2006).

[6]   Ali, A., Wenk, M. R. & Lear, M. J., Total synthesis of a fully lipidated form of phosphatidylinositol dimanoside (PIM-2) of Mycobacterium tuberculosis, Tetrahedron Lett 50, 5664-5666 (2009).

[7]   Boonyarattanakalin, S. et al. Chemical Synthesis of All Phosphatidylinositol Mannoside (PIM) Glycans from Mycobacterium tuberculosis. J Am Chem Soc 130, 16791-16799 (2008).

[8]   Kamena, F. et al. Synthetic GPI array to study antitoxic malaria response. Nat Chem Biol 4, 238-240 (2008).

[9]   Schofield, L. et al. Synthetic GPI as a candidate antitoxic vaccine in a model of malaria. Nature 418, 785-789 (2002).

[10] Fraser-Reid, B. et al. Efficient Chemical Synthesis of a Dodecasaccharidyl Lipomannan Component of Mycobacterial Lipoarabinomannan (dagger). J Org Chem 73, (24), 9732-9743 (2008)

[11] Jayaprakash, K. N. et al. Synthesis of a lipomannan component of the cell-wall complex of Mycobacterium tuberculosis is based on Paulsen's concept of donor/acceptor "match". Angew Chem Int Ed Eng 44, (36), 5894-5898 (2005).

[12] Lu, J. et al. A strategy for ready preparation of glycolipids for multivalent presentation. Org Lett 7, (18), 3841-3843 (2005).

[13] Buskas, T. et al. Immunotherapy for cancer: synthetic carbohydrate-based vaccines. Chem Commun 36, 5335-5349 (2009).

[14] Kunz, W. U. et al. Synthetic Vaccines Consisting of Tumor-Associated MUC1 Glycopeptide Antigens and a T-Cell Epitope for the Induction of a Highly Specific Humoral Immune Response. Angew Chem Int Ed 47, 7551-7556 (2008).

[15] Zhu, X. & Schmidt, R. R. New Principles for Glycoside-Bond Formation. Angew. Chem Int Ed 48, 1900-1934 (2009).

[16] Schmidt, R. R. et al. New Aspects of glycoside bond formation. Pure Appl Chem 71, 729-744 (1999).

[17] Zhu, J., Wan, Q. et al. Biologics through Chemistry: Total Synthesis of a Proposed Dual-Acting Vaccine Targeting Ovarian Cancer by Orchestration of Oligosaccharide and Polypeptide Domains. J Am Chem Soc 131, 4151-4158 (2009)

[18] Zhu, J., Wan, Q. et al. From Synthesis to Biologics: Preclinical Data on a Chemistry Derived Anticancer Vaccine. J. Am. Chem Soc 131, 9298-9303 (2009).

[19] Wilson, R. M. & Danishefsky, S. J. Promising agents at the interface of biology and oncology derived through chemical synthesis. Pure Appl Chem 79, (12), 2189-2216 (2007).

[20] Seeberger, P. H. Chemical glycobiology: why now? Nat Chem Biol 3, 368-372 (2009).

[21] Hecht, M. L et al. Recent advances in carbohydrate-based vaccines. Curr Opin Chem Biol 13, (3), 354-359 (2009).

[22] Seeberger, P. H. Automated carbohydrate synthesis as platform to address fundamental aspects of glycobiology - current status and future challenges. Carbohydr Res 343, (12), 1889-1896 (2008)

[23] Horlacher, T. & Seeberger, P.H. Carbohydrate arrays as tools for research and diagnostics. Chem. Soc. Rev. 37, (7), 1414-1422 (2008).

[24] Agard, N. J. & Bertozzi, C. R. Chemical approaches to perturb, profile, and perceive glycans. Acc Chem Res 42, (6), 788-797 (2009).

[25] Laughlin, S. T. & Bertozzi, C. R. Imaging the glycome. Proc Nat Acad Sci USA 106, (1), 12-17 (2009).

[26] Gamblin, D. P. et al. Glycoprotein synthesis: an update. Chem Rev 109, (1), 131-163 (2009).

[27] Davis, B.G. The linear assembly of a pure glycoenzyme. Angew Chemie Int Ed 48, (26), 4674-4678 (2009).

[28] Wu, C.-Y. et al. New development of glycan arrays. Org Biomol Chem 7, (11), 2247-54 (2009).

[29] Wu, D. et al. Glycolipids as immunostimulating agents. Bioorg Med Chem 16, (3), 1073-83 (2008).

[30] Galonic, D. P. & Gin, D. Y. Chemical glycosylation in the synthesis of glycoconjugate antitumour vaccines. Nature 446, (7139), 1000-1007 (2007).

[31] Deng, K. et al. Synthesis and structure verification of the vaccine adjuvant QS-7-Api. Synthetic access to homogeneous Quillaja saponaria immunostimulants. J Am Chem Soc 130, (18), 5860-5861 (2008)

[32] Ferguson, M. A. et al. Glycosyl-phosphatidylinositol moiety that anchors Trypanosoma brucei variant surface glycoprotein to the membrane. Science 239, 753-759 (1988).

[33] Low, M. GPI anchored biomolecules-an overview, in GPI anchored membrane proteins and carbohydrates (Hoessli, D. C.,and Ilangumaran, S., Eds.) pp 1-14, Landes company, Austin, Tx. (1999).

[34] Nosjean, O. et al. Mammalian GPI proteins: sorting, membrane residence and functions. Biochim Biophys Acta 1331, 153-186 (1997).

[35] Thomas, J. R. et al. Structure, Biosynthesis and Function of glycosylphosphatidylinositols. Biochemistry 29, 5413-5422 (1990).

[36] Ruhela. D. & Vishwakarma R. A. Iterative Synthesis of Leishmania Phosphoglycans by Solution,Solid-Phase, and Polycondensation Approaches without involving any Glycosylation. J Org Chem 68, 4446-4456 (2003).

[37] Low, M. G. Glycosyl-phosphatidylinositol: a versatile anchor for cell surface proteins. FASEB J 3, 1600-1608 (1989).

[38] Low, M. G. & Saltiel, A. R. Structural and functional roles of glycosyl-phosphatidylinositol in membranes. Science 239, 268-275 (1988).

[39] Turco, S. J. & Descoteaux, A. The Lipophosphoglycan of Leishmania parasites. Ann Rev Microbiol 46, 65-94 (1992).

[40] McConville, M. J. & Ferguson, M. A. J. The structure, biosynthesis and function of glycosylated phosphatidylinositols in the parasitic protozoa and higher eukaryotes. Biochem J 294, 305-324 (1993).

[41] Ferguson, M. A. J. The structure, biosynthesis and functions of glycosylphosphatidylinositol anchors, and the contributions of trypanosome research. J Cell Sci 112, 2799-2809 (1999).

[42] Ferguson, M. A. J. et al. Glycosylsn-1,2-dimyristylphosphatidylinositol is covalently linked to Trypanosoma brucei variant surface glycoprotein. J Biol Chem 260, 14547-14555 (1985).

[43] Homans, S. W. et al. Complete structure of the glycosyl phosphatidylinositol membrane anchor of rat brain Thy-1 glycoprotein. Nature 333, 269-272 (1988).

[44] Eisenhaber, B. et al. Enzymes and auxiliary factors for GPI lipid anchor biosynthesis and post-translational transfer to proteins. BioEssays 25, 367-385 (2003).

[45] Tiede, A. et al. Biosynthesis of glycosylphosphatidylinositols in mammals and unicellular microbes. Biol Chem 380, 503-523 (1999).

[46] Chesebro, B. et al. Anchorless prion protein results in infectious amyloid disease without clinical scrapie. Science 308, 1435-1439 (2005).

[47] Kawagoe, K. et al. Glycosylphosphatidylinositol- anchor-deficient mice: implications for clonal dominance of mutant cells in paroxysmal nocturnal hemoglobinuria. Blood 87, 3600-3606 (1996).

[48] Leidich, S. D. et al. A conditionally lethal yeast mutant blocked at the first step in glycosyl phosphatidylinositols anchor synthesis. J Biol Chem 269, 10193-10196 (1994).

[49] Deeg, M. A. et al. Glycan components in the glycoinositol phospholipid anchor of human erythrocyte acetylcholinesterase. Novel fragments produced by trifluoroacetic acid. J Biol Chem 267, 18573-18580 (1992).

[50] Brewis, I. A. et al. Structures of the glycosyl-phosphatidylinositol anchors of porcine and human renal membrane dipeptidase. Comprehensive structural studies on the porcine anchor and interspecies comparison of the glycan core structures. J Biol Chem 270, 22946-22956 (1995).

[51] Nakano, Y. et al. Structural study on the glycosyl-phosphatidylinositol anchor and the asparagine-linked sugar chain of a soluble form of CD59 in human urine. Arch Biochem Biophys 311, 117-126 (1994).

[52] Mukasa, R. et al. Characterization of glycosylphosphatidylinositol (GPI)-anchored NCAM on mouse skeletal muscle cell line C2C12: the structure of the GPI glycan and release during myogenesis. Arch Biochem Biophys 318, 182-190 (1995).

[53] Fontaine, T. et al. Structures of the glycosylphosphatidylinositol membrane anchors from Aspergillus fumigatus membrane proteins. Glycobiology 13, 169-177 (2003).

[54] Oxley, D. & Bacic, A. Structure of the glycosylphosphatidylinositol anchor of an arabinogalactan protein from Pyrus communis suspension-cultured cells. Proc Natl Acad Sci USA 96, 14246-14251 (1999).

[55] Macrae, J. I. et al. Structural characterization of NETNES, a novel glycoconjugate in Trypanosoma cruzi epimastigotes. J Biol Chem 280, 12201-12211 (2005).

[56] Frankhauser, C. et al. Structures of Glycosylphosphatidylinositol Membrane Anchors from Saccharomyces cerevisiae. J Biol Chem 268, 26365-26374 (1993).

[57] Striepen, B. et al. Molecular structure of the "Low Molecular Weight Antigen" of Toxoplasma gondii: A Glucose α1-4 N-Acetylgalactosamine makes free Glycosyl-Phosphatidylinositols highly immunogenic. J Mol Biol 266, 797-813 (1997).

[58] McConville, M. J. & Ferguson, M. A. J. The structure, biosynthesis and function of glycosylated phosphatidylinositols in the parasitic protozoa and higher eukaryotes. Biochem J 294, 305-324 (1993).

[59]   Schroter, S. et al. Male-specific Modification of Human CD52. J Biol Chem 274, 29862-29873 (1999).

[60]   Treumann, A. et al. Primary structure of CD52. J Biol Chem 270, 6088-6099 (1995).

[61]   Gerold, P. et al. Glycosylphosphatidylinositols synthesized by Asexual Erythrocytic stages of the Malarial Parasite, Plasmodium falciparum. J Biol Chem 269, 2597-2606 (1994).

[62]   Gerold, P. et al. Structural anchor of analysis of the glycosyl-phosphatidylinositol membrane the merozoite surface proteins-l and -2 of Plasmodium falciparum. Mol Biochem Parasitol 75, 131-143 (1996).

[63]   Gowda, D. C. et al. Glycosylphosphatidylinositol Anchors Represent the Major Carbohydrate Modification in Proteins of Intraerythrocytic Stage Plasmodium falciparum. J Biol Chem 272, 6428-6439 (1997).

[64]   Stahl, N. et al. Glycosylinositol Phospholipid Anchors of the Scrapie and Cellular Prion Proteins contain Sialic Acid. Biochemistry 31, 5043 (1992).

[65]   MacRae, J. I. et al. Structural Characterization of NETNES, a Novel Glycoconjugate in Trypanosoma cruzi Epimastigotes. J Biol Chem 280, 12201-12211 (2005).

[66]   Maier, M. E. Structural revisions of natural products by total synthesis. Nat Prod Rep 26, 1105-1124 (2009).

[67]   Nicolaou, K. C. & Snyder, S. A. Chasing Molecules That Were Never There: Misassigned Natural Products and the Role of Chemical Synthesis in Modern Structure Elucidation. Angew Chem Int Ed 44, 1012-1044 (2005).

[68]   Baeschlin, D. K. et al. Rapid assembly of oligosaccharides: Total synthesis of a glycosylphosphatidylinositol anchor of Trypanosoma brucei. Angew Chem Int Ed 37, 3423-3428 (1999).

[69]   Baeschlin, D. K. et al. 1,2-Diacetals in synthesis : Total synthesis of a glycosylphosphatidylinositol anchor of Trypanosoma brucei. Chem Eur J 6, 172-186 (2000).

[70]   Ley, S. V. et al. Dispiroketals in Synthesis (Part 22)1: Use of Chiral 2,2'-Bis(phenylthiomethyl)dihydropyrans as new protecting and resolving agents for 1,2-Diols. Synlett 791-792 (1996).

[71]   Watanabe, Y. et al. Highly efficient protection by the Tetraisopropyldisiloxane-l ,3-diyl Group in the synthesis of myo-lnositol Phosphates as lnositol 1,3,4,6-Tetrakisphosphate. J Chem Soc Chem Commun 482-483 (1989).

[72]   Ley, S. V. et al. Dispiroketals in Synthesis (Part 20)[1]: Preparation of Chiral 2,2'-Bis(halomethyl) and 2,2'-Bis(phenylthiomethyl)dihydropyrans. Synlett 787-788 (1996).

[73]   Murakata, C. & Ogawa, T. Synthetic studies on cell-surface glycans 85. Stereoselective total synthesis of the glycosyl phosphatidylinositols (GPI) anchor of Trypanosoma brucei. Carbohydr Res 235, 95-114 (1992).

[74]   Murakata, C. & Ogawa, T. Synthetic studies on cell-surface glycans. 78. A total synthesis of GPI anchor of Trypanosoma brucei. Tetrahedron. Lett 32, 671-674 (1991).

[75]   Baeschlin D. K. et al. 1,2-Diacetals in synthesis : Total synthesis of glycosylphosphatidylinositol anchor of Trypanosoma brucei. Chemistry (Weinheim an der Bergstrasse, Germany) 6, 172-86 (2000).

[76]   Khiar, N. & Martin-Lomas, M. A Highly Convergent Synthesis of the Tetragalactose Moiety of the GPI Anchor of the VSG of Trypanosoma brucei. J Org Chem 60, 7017-21 (1995).

[77]   Murakata, C. & Ogawa, T. Synthetic studies on glycosyl-phosphatidylinositol anchor of Trypanosoma brucei. Tennen Yuki Kagobutsu Toronkai Koen Yoshishu 33rd 47-53 (1991).

[78]   Ziegler, T. et al. Synthesis of alpha-galactosylated fragments related to the core-structure of the GPI anchor of Trypanosoma brucei. Carbohydr Res 327, 367-75 (2000).

[79]   Liu, X. et al. Convergent Synthesis of a Fully Lipidated Glycosylphosphatidylinositol anchor of Plasmodium falciparum. J Am Chem Soc 127, 5004-5005 (2005).

[80]   Schofield, L. et al. Neutralizing monoclonal antibodies to glycosylphosphatidylinositol, the dominant TNF-alpha-inducing toxin of Plasmodium falciparum: prospects for the immunotherapy of severe malaria. Annu Trop Med Parasitol 87, 617-626 (1993).

[81]   Schofield, L. et al. Synthetic GPI as a candidate antitoxic vaccine in a model of malaria. Nature 418, 785-788 (2002).

[82]   Seeberger, P. H. et al. A convergent, versatile route to two synthetic conjugate anti-toxin malaria vaccines. Chem Commun 1706-1707 (2004).

[83]   Lu, J. et al. Synthesis of a Malaria Candidate Glycosylphosphatidylinositol (GPI) Structure: A Strategy for Fully Inositol Acylated and Phosphorylated GPIs. J Am Chem Soc 126, 7540-7547 (2004).

[84]   Lopez-Prados, J. & Martin-Lomas, M. Inositolphosphoglycan mediators: An effective synthesis of the conserved linear GPI anchor structure. J Carbohydr Chem 24, 393-414 (2005).

[85]   Kwon, Y-Uk. et al. Assembly of a series of malarial glycosylphosphatidylinositol anchor oligosaccharides. Chem Eur J 11, 2493-2504 (2005).

[86] Lu, J. et al. First synthesis of a malarial prototype: a fully lipidated and phosphorylated GPI membrane anchor. Tetrahedron Lett 45, 879-882 (2004).

[87] Hewitt. M. C. et al. Rapid synthesis of a Glycosylphosphatidylinositol based Malaria Vaccine using automated solid-phase oligosaccharide Synthesis. J Am Chem Soc 124, 13434-13436 (2002).

[88] Becker, C. F. W. et al. Semisynthesis of a glycosylphosphatidylinositol-anchored prion protein. Angew Chem Int Ed 47, 8215-8219 (2008).

[89] Pekari. K. & Schmidt. R.R. A variable concept for the preparation of branched glycosyl phosphatidyl inositol anchors. J Org Chem 68, 1295-308 (2003).

[90] Pekari, K. et al. Synthesis of the Fully Phosphorylated GPI Anchor Pseudohexasaccharide of Toxoplasma gondii J Org Chem 66, 7432-7442 (2001).

[91] Udodong, U.E. et al. A ready, convergent synthesis of the heptasaccharide GPI membrane anchor of rat brain Thy-1 glycoprotein. J Am Chem Soc 115, 17, 7886-7887 (1993).

[92] Campbell, A. S. & Fraser-Reid, B. First synthesis of a fully Phosphorylated GPI Membrane Anchor: Rat Brain Thy-1. J Am Chem Soc 117, 41, 10387-10388 (1995).

[93] Campbell, A. S. & Fraser-Reid, B. Support Studies for Installing the Phosphodiester Residues of the Thy-1 Glycoprotein Membrane Anchor. Bioorg Med Chem 2, 1209-1219 (1994).

[94] James, L. C. et al. 1.9 Å Structure of the therapeutic Antibody CAMPATH-1H Fab in complex with a synthetic peptide Antigen. J Mol Biol 289, 293-301 (1999).

[95] Calne, R. et al. Prope tolerance, perioperative CAMPATH-1H, and low-dose cyclosporin monotherapy in renal allograft recipients. Lancet 351, 1701-1702 (1998).

[96] Pawson, R. et al. Treatment of T-cell prolymphocytic leukemia with human CD52 antibody. J Clin Oncol 15, 2667-2672 (1997).

[97] Osterborg, A. et al. Clonal CD8$^+$ and CD52$^-$ T cells are induced in responding B cell lymphoma patients treated with CAMPATH-1H* (anti-CD52). Eur J Haematol 58, 5-13 (1997).

[98] Moreau, T. et al. Preliminary evidence from magnetic resonance imaging for reduction in disease activity after lymphocyte depletion in multiple sclerosis. Lancet 344, 298-301 (1994).

[99] Mathieson, P. W. et al. Monoclonal-antibody therapy in systemic vasculitis. New Engl J Med 323, 250-254 (1990).

[100] Isaac, J. D. et al. Humanized monoclonal antibody therapy for rheumatoid arthritis. Lancet 340, 748-752 (1992).

[101] Tsuji, Y. Carbohydrate antigens recognized by anti-sperm antibodies; Kurpisz, M., Fernandez, N., Eds.; BIOS Sci. Pub.: Oxford, 1995; pp 23-32.

[102] Tsuji, Y. et al. Human sperm carbohydrate antigens defined by an antisperm human monoclonal antibody derived from an infertile woman bearing antisperm antibodies in her serum. J Exp Med 168, 343-356 (1988).

[103] Wu, X. & Guo, Z. Convergent Synthesis of a Fully Phosphorylated anchor of the CD52 Antigen. Org Lett 9, 4311-4313 (2007).

[104] Xue, J. et al. First Total Synthesis of a GPI anchored Peptide. J. Org Chem 68, 4020-4029 (2003).

[105] Xue, J. & Guo, Z. Convergent synthesis of an inner core GPI of sperm CD52. Bioorg. Med. Chem. Lett 12, 2015-2018 (2002).

[106] Lemieux, R. U. et al. Halide Ion Catalyzed Glycosidation Reactions. Syntheses of a-Linked Disaccharides. J Am Chem Soc 97, 4056-4602 (1975).

[107] Xue, J. & Guo, Z. Convergent synthesis of a GPI containing an acylated Inositol. J Am Chem Soc 125, 16334 (2003).

[108] Wu, X. et al . Synthesis and biological evaluation of Sperm CD52 GPI anchor and related derivatives as binding receptors of pore-forming CAMP factor. Carbohydr Res 343, 1718-1729 (2008).

[109] Wu, X. et al. Synthesis and biological evaluation of sperm CD52 GPI anchor and related derivatives as binding receptors of pore-forming CAMP factor. Carbohydr. Res 344, 952 (2009).

[110] Hederos, M. & Konradsson, P. Synthesis of the Trypanosoma cruzi LPPG Heptasaccharyl myo-Inositol. J Am Chem Soc 128, 3414-3419 (2006).

[111] Lindberg, J. et al. Efficient routes to glucosamine-myo-inositol derivatives, key building blocks in the synthesis of glycosylphosphatidylinositol anchor substances. Tetrahedron 58, 1387-1398 (2002).

[112] Lahmann, M. et al. Synthesis of a polyphosphorylated GPI - anchor core structure. Can J Chem 80, 1105-1111 (2002).

[113] Mayer, T.G. et al. Synthesis of a GPI anchor of the yeast Saccharomyces cerevisiae. Angew Chem Intd Ed 106, 2289-2293 (1994).

[114] Ma, Z. et al. Synthesis of two oligosaccharide, the GPI anchor glycans from S. cerevisiae and A. fumigatus. Carbohydr Res 339, 29-35 (2004).

[115] Vishwakarma, R. A. & Menon, A. K. Flip-flop of glycosylphosphatidylinositols (GPI's) across the ER. Chem Comm 453-455 (2005).

[116] Eisenhaber, B. et al. Enzymes and auxiliary factors for GPI lipid anchor biosynthesis and post-translational transfer to proteins. BioEssays 25, 367-385 (2003).

[117] Kinoshita, T. & Inoue, N. Dissecting and manipulating the pathway for glycosylphosphatidylinositol-anchor biosynthesis. Curr Opin Chem Biol 4, 632-638 (2000).

[118] Tiede, A. et al. Biosynthesis of glycosylphosphatidylinositols in mammals and unicellular microbes. Biol Chem 380, 503- 23 (1999).

[119] Hirose, S. et al. Derivation and characterization of glycoinositol-phospholipid anchor- defective human K562 cell clones. J Biol Chem 267, 5272-5278 (1992).

[120] Bastisch, I. et al. Glycosylphosphatidylinositol (GPI)-deficient Jurkat T cells as a model to study functions of GPI anchored proteins. J Clin Exp Immunol 122, 49-54 (2000).

[121] Kawagoe, K. et al. Glycosylphosphatidylinositol -Anchor - Deficient Mice: Implications for Clonal Dominance of Mutant Cells in Paroxysmal Nocturnal Hemoglobinuria. Blood 87, 3600-3606 (1996).

[122] Paulick, M. G. & Bertozzi, C. R. The glycosylphosphatidylinositol Anchor: A Complex Membrane-Anchoring Structure for Protein. Biochemistry 47, 6991-7000 (2008).

[123] Pegoraro, S. & Moroder, L. in Houben-Weyl Methods of Organic Chemistry, Synthesis of Peptides and Peptidomimetics, Vol.E 22b (Eds.: M. Goodman, A. Felix, L. Moroder, C. Toniolo), Thieme, Stuttgart, pp. 333-374 (2003).

[124] Muir, T. W. Semisynthesis of Proteins by expressed protein ligation. Annu Rev Biochem 72, 249-289 (2003).

[125] Dawson, P. E. & Kent, S. B. H. Synthesis of Native Proteins by Chemical Ligation. Annu Rev Biochem 69, 923-960 (2000).

[126] Casi, G. & Hilvert, D. Convergent protein synthesis. Curr Opin Struct Biol 13, 589-594 (2003).

[127] David, R., Richter, M. P. O. & Beck-Sickinger, A. Expressed protein ligation Method and applications. Eur J Biochem 271, 663-677 (2004).

[128] Song, H. Y. et al. Practical synthesis of maleimides and coumarin-linked probes for protein and antibody labelling via reduction of native disulfides, Org Biomol Chem 7, 3400-3406 (2009).

[129] R. Huisgen in 1,3-Dipolar Cycloaddition Chemistry, (Ed.: A. Padwa), Wiley, New York, 1984, pp. 1-176.

[130] Rostovtsev, V. V. et al. A Stepwise Huisgen Cycloaddition Process:Copper(I)-Catalyzed Regioselective "Ligation" of Azides and Terminal Alkynes. Angew Chem Int Ed 41, 2596-2599 (2002).

[131] Tornøe, C. W. et al. Peptidotriazoles on Solid Phase: [1,2,3]-Triazoles by Regiospecific Copper(I)-Catalyzed 1,3-Dipolar Cycloadditions of Terminal Alkynes to Azides. J Org Chem 67, 3057-3064 (2002).

[132] Musiol, H-J. et al. Toward Semisynthetic Lipoproteins by Convergent Strategies Based on Click and Ligation Chemistry. ChemBioChem 6, 625-628 (2005).

[133] Yang, L. et al. Integral membrane proteins of the nuclear envelope are dispersed throughout the endoplasmic reticulum during mitosis. J Cell Biol 137, 1199-1210 (1997).

[134] Terasaki, M. & Reese, T. S. Characterization of endoplasmic reticulum by co-localization of BiP and dicarbocyanine dyes. J Cell Sci 101, 315-322 (1992).

[135] Endo, T. et al. Diversity of oligosaccharide structures linked to asparagines of the scrapie prion protein. Biochemistry 28, 8380-8388 (1989).

[136] Stahl, N. et al. Scrapie prion protein contains a phosphatidylinositol glycolipid. Cell 51, 229-240 (1987).

[137] Breydo, L. et al. Nonpolar Substitution at the C-Terminus of the Prion Protein, a Mimic of the Glycosylphosphatidylinositol Anchor, Partially Impairs Amyloid Fibril Formation. Biochemistry 46, 852-861 (2007).

[138] Legname, G. et al. Synthetic mammalian prions. Science 305, 673-676 (2004).

[139] Baskakov, I. V. et al. Pathway complexity of prion protein assembly into amyloid. J Biol Chem 277, 21140-21148 (2002).

[140] Olschewski, D. et al. Semisynthetic Murine Prion Protein Equipped with a GPI Anchor Mimic Incorporates into Cellular Membranes. Chem Biol 14, 994-1006 (2007).

[141] Rose, K. & Vizzavona, J. Stepwise solid-phase synthesis of polyamides as linkers. J Am Chem Soc 121, 7034-7038 (1999).

[142] Becker, C.F. et al. Chemical synthesis and single channel properties of tetrameric and pentameric TASPs (template-assembled-synthetic proteins) derived from the transmembrane domain of HIV virus protein u (Vpu). J Biol Chem 279, 17483-17489 (2004).

[143] Shin, Y. et al. Fmoc-Based Synthesis of Peptide-RThioesters: Application to the Total Chemical Synthesis of a Glycoprotein by Native Chemical Ligation. J Am Chem Soc 121, 11684-11689 (1999).

[144] Paulick, M. G. et al. A chemical approach to unraveling the biological function of the glycosylphosphatidylinositol anchor. Proc Natl Acad Sci U.S.A. 104, 20332-20337 (2007).

[145] Grogan, M. J. et al. Synthesis of lipidated green fluorescent protein and its incorporation in supported lipid bilayers. J Am Chem Soc 127, 14383-14387 (2005).

[146] Van den Berg, C.W. et al. Exogenous Glycosyl Phosphatidylinositol-anchored CD59 Associates with Kinases in Membrane Clusters on U937 Cells and Becomes Ca^{2+}-signaling Competent. J Cell Biol 131, 669-677 (1995).

[147] Medof, M. E. et al. Inhibition of complement activation on the surface of cells after incorporation of Decay-accelerating factor (DAF) into the membranes. J Exp Med 160, 1558-1578 (1984).

[148] Premkumar, D. R. D. et al. Properties of Exogenously Added GPI anchored Proteins Following Their Incorporation Into Cells. J Cell Biochem 82, 234-245 (2001).

[149] Medof, M. E. et al. Cell surface engineering with GPI anchored Proteins. FASEB. J 10, 574-586 (1996).

[150] Laughlin, S. T. & Bertozzi, C. R. Imaging the glycome. Proc Natl Acad Sci USA 106, 12-17 (2009).

[151] Saxon, E. & Bertozzi, C. R. Cell surface engineering by a modified Staudinger reaction. Science 287, 2007-2010 (2000).

[152] Agnew, H. D. et al. Iterative In Situ Click Chemistry Creates Antibody-like Protein-Capture Agents. Angew Chem Intd Ed 48, 4944-4948 (2009).

[153] Whiting, M. et al. Inhibitors of HIV-1 Protease by Using In Situ Click Chemistry. Angew Chem Intd Ed 45, 1435-1439 (2006).

[154] Agard, N. J. et al. A strain-promoted [3 + 2] azide-alkyne cycloaddition for covalent modification of biomolecules in living systems. J Am Chem Soc 126, 15046-15047 (2004).

[155] Baskin, J. M. Copper-free click chemistry for dynamic in vivo imaging. Proc Natl Acad Sci USA 104, 16793-16797 (2007).

[156] Agard, N. J. et al. A comparative study of bioorthogonal reactions with azides. ACS Chem Biol 1, 644-648 (2006).

[157] Luchansky, S. J. et al. Expanding the diversity of unnatural cell surface sialic acids. Chem Bio Chem 5, 371-374 (2004).

[158] Kosa, R. E. et al. Modification of cell surfaces by enzymatic introduction of special sialic acid analogues. Biochem Biophys Res Commun 190, 914-920 (1993).

[159] Hsu, T. L. et al. Alkynyl sugar analogs for the labeling and visualization of glycoconjugates in cells. Proc Natl Acad Sci U.S.A. 104, 2614-2619 (2007).

[160] Dube, D. H. et al. Probing mucin-type O-linked glycosylation in living animals. Proc Natl Acad Sci USA 103, 4819-4824 (2006).

[161] Laughlin, S. T. et al. In vivo imaging of membrane associated glycans in developing zebrafish. Science 320, 664-667 (2008).

[162] Hart, G. W. et al. Cycling of O-linked beta-N-acetylglucosamine on nucleocytoplasmic proteins. Nature 446, 1017-1022 (2007).

[163] Rabuka, D. et al. A chemical reporter strategy to probe glycoprotein fucosylation. J Am Chem Soc 128, 12078-12079 (2006).

[164] Sawa, M. et al. Glycoproteomic probes for fluorescent imaging of fucosylated glycans in vivo. Proc Natl Acad Sci U.S.A. 103, 12371-12376 (2006).

[165] Lemieux, G. A. et al. A fluorogenic dye activated by the staudinger ligation. J Am Chem Soc 125, 4708-4709 (2003).

[166] Hangauer, M. J. & Bertozzi, C. R. A FRET-based fluorogenic phosphine for live-cell imaging with the Staudinger ligation. Angew Chem Int Ed 47, 2394-2397 (2008).

[167] Hmama, Z. & Av-Gay, Y. Mycobacterial manipulation of the host cell. FEMS Microbiol Rev 29, 1041-1050 (2005).

[168] Seeberger, P. H. Automated carbohydrate synthesis as platform to address fundamental aspects of glycobiology—current status and future challenges. Carbohydr Res 343, 1889-1896 (2008).

[169] Nigou, M. et al. Lipoarabinomannans: from structure to biosynthesis. Biochimie 85, 153-166 (2003).

[170] Apostolou, I. et al. Murine natural killer cells contribute to the granulomatous reaction caused by mycobacterial cell walls. Proc Natl Acad Sci U.S.A. 96, 5141-5146 (1999).

[171] Raviglione, M. C. & Uplekar, M. W. WHO's new Stop TB Strategy. Lancet 367, 952-955 (2006).

[172] Adams, E. W. et al. Oligosaccharide and Glycoprotein Microarrays as Tools in HIV Glycobiology: Glycan-Dependent gp120/Protein Interactions. Chem Biol 11, 875 (2004).

[173] Kwon, Y. et al. Assembly of a Series of Malarial Glycosylphosphatidylinositol Anchor Oligosaccharides. Chem Eur J 11, 2493-2504 (2005).

[174] Bochner, B. S. et al. Glycan Array Screening Reveals a Candidate Ligand for Siglec-8. J Biol Chem 280, 4307 (2005).

[175] Bryan, M. C. et al. High-throughput identification of fucosyltransferase inhibitors using carbohydrate microarrays. Bioorg Med Chem Lett 14, 3185 (2004).

[176] Park, S. & Shin, I. Carbohydrate Microarrays for Assaying Galactosyltransferase Activity. Org Lett 9, 1675 (2007).

[177] Blixt, O. et al. Glycan microarrays for screening sialyltransferase specificities. Glycoconjugate J 25, 59 (2008).

[178] Noti, C. et al. Preparation and Use of Microarrays Containing Synthetic Heparin Oligosaccharides for the Rapid Analysis of Heparin-Protein Interactions. Chem Eur J 12, 8664 (2006).

[179] Shipp, E. L. & Hsieh-Wilson, L. C. Profiling the Sulfation Specificities of Glycosaminoglycan Interactions with Growth Factors and Chemotactic Proteins Using Microarrays. Chem Biol 14, 195 (2007).

[180] Gama, C. I. et al. Sulfation patterns of glycosaminoglycans encode molecular recognition and activity. Nat Chem Biol 2, 467 (2006).

[181] Disney, M. D. & Seeberger, P. H. Aminoglycoside Microarrays to Explore Interactions of Antibiotics with RNAs and Proteins. Chem Eur J 10, 3308 (2004).

[182] Stevens, J. et al. Glycan microarray technologies: tools to survey host specificity of influenza viruses. Nat Rev Microbiol 4, 857 (2006).

[183] Harland, C. W. et al. Synthetic Trehalose Glycolipids Confer Desiccation Resistance to Supported Lipid Monolayers. Langmuir 25, 5193-5198 (2009).

[184] Agu, C. A. et al. The cytotoxic activity of the bacteriophage lambda-holin protein reduces tumour growth rates in mammary cancer cell xenograft models. J Gene Med 8, 229-241 (2006).

[185] Metzner, C. et al. Association of glycosylphosphatidylinositol-anchored protein with retroviral particles. FASEB J 22, (8), 2734-2739 (2008)

[186] Metzner, C. et al. Rafts, anchors and viruses-a role for glycosylphosphatidylinositol anchored proteins in the modification of enveloped viruses and viral vectors. Virology 382, 125-131 (2008).

[187] Lear, M. J. et al. Singapore R&D and globetrotting. Biotech J 4, (2), 179-185 (2009).

[188] Lear M. J. & Reux B., unpublished results.

CHAPTER 6

Surface Engineering of Biomembranes with GPI-Anchored Proteins and its Applications

Christoph Metzner, Daniel F. Legler and John A. Dangerfield

Abstract: This chapter covers the use of glycosylphosphatidylinositol (GPI)-anchored proteins for surface modification of diverse types of biomembrane covered entities ranging from viruses and virus-related particles (section 1), to cells (section 2) and other natural and engineered micro- and nano-scaled particles (section 3). The aim is to present and review state-of-the-art research in this area and to discuss the future direction of GPI painting technology relating to applications in research, biotechnology and biomedicine.

GPI-ANCHORED PROTEIN MODIFICATION OF VIRUSES, VIRAL VECTORS AND VIRUS-LIKE PARTICLES

Background

Viruses do not only cause a large range of pathological conditions but are also used as templates for the generation of biotechnological devices such as viral vectors (VVs) and virus-like particles (VLPs) and commonly used systems derive from adeno-, retro- and lentiviral species. VVs are genetically modified virus particles designed to deliver specific genetic information to cells. This is especially useful in a research setting, when foreign proteins need to be expressed long-term in eukaryotic cells or in gene therapy, where foreign DNA is substituting incomplete or inactive endogenous copies. VLPs contain the structural elements of viruses; however the viral genome and/or functional properties are partly or completely absent. Thus, VLPs cannot infect cells like the viruses or VVs. VLPs are used as particulate carrier systems to provide *e.g.* strong immunogens in vaccine development [1] or a more localised distribution of otherwise soluble agents [2].

Viruses are grouped into two categories based on their structural composition. Enveloped viruses (EV) consist of the genetic material encased in a protein derived core which is in turn surrounded by a lipid bilayer membrane – the envelope. „Naked" viruses lack such envelope-structures (see Table **1** for a list of enveloped virus families). Presence or absence of envelope structures strongly influence the infectious cycle of viruses and as a consequence therapy strategies differ. One interesting element of virus biology is the apparent association of both enveloped [3-18] (Table **1**) and non-enveloped viruses [19-22] with specific membrane microdomains or lipid rafts.

Although the concept of membrane microdomains and membrane heterogeneity is widely accepted now, the details remain elusive [23-26]. They are proposed as sites for entry and exit of a number of different virus species (for reviews see [27-29]). Furthermore, proteins post-translationally modified with a glycosylphosphatidylinositol-(GPI)-anchor are also not only predominantly localised at the cell membrane but enriched in membrane microdomains. Consequentially, enveloped viral particles can incorporate GPI-anchored proteins when exiting from cells. This has been specifically studied in retroviruses. For example, human immunodeficiency virus 1 (HIV-1) particles carry cell membrane-derived molecules of GPI-anchored CD59 and CD55 after exit from cells [30-32]. This mechanism is beneficial for the virus because CD55 and CD59 inhibit complement activity and can thus protect viral particles from this branch of innate immunity [30-32]. Another important aspect of the biology of GPI-anchored proteins in this context is the capability of purified extracts to re-insert into lipid membranes. This process is mediated by the lipid residues of the GPI-anchor. The re-insertion occurs not only into eukaryotic cells [33,34], but also into the envelopes of viral particles [35] and was described for native GPI-anchored proteins, as well as for proteins that acquired a GPI anchor following genetic modification [36,37]. This phenomenon, termed "painting", together with the co-localisation of enveloped virus exit and GPI-proteins at membrane microdomains provides the framework for the surface modification of enveloped viruses as described here (Fig. **1**). Both aspects require the presence of lipid structures surrounding the virus particles, thus only enveloped viruses can be modified.

Table 1: Enveloped viruses and membrane microdomain-associated exit. The table shows the families of enveloped viruses with relevant examples, which may be modified by GPI-anchored proteins. Crosses indicate association of the viral budding processes with membrane microdomain structures. References pertaining to these are given.

FAMILY	EXAMPLE	LR/EXIT	REFERENCE
Herpesviridae	Epstein-Barr virus (EBV)		
	Herpes simplex virus type 1 (HSV-1)		
Hepadnaviridae	Hepatitis B virus		
Poxviridae	Vaccinia virus		
Togaviridae	Rubella virus		
Arenaviridae	lymphocytic choriomeningitisvirus		
Flaviviridae	West-Nile virus		
	Dengue virus		
Orthomyxoviridae	Influenza virus	X	3-5
Paramyxoviridae	Measles virus	X	6.7
	Respiratory syncytial virus (RSV)	X	8-10
	Sendai virus	X	11
Bunyaviridae	Hanta virus		
Rhabdoviridae	Vesicular stomatitis virus (VSV)	X	12
	Rabies virus		
Filoviridae	Marburg virus	X	13,14
	Ebola virus	X	13-15
Coronaviridae	severe acute respiratory syndrome (SARS) associated Corona virus		
Bornaviridae	Borna disease virus		
Retroviridae	Human immunodeficiency virus (HIV-1)	X	16,17
	Murine leukemia virus (MLV)	X	18

State of the Art

Modification of Viral Particles during Cell Exit

GPI-anchored proteins are enriched in membrane microdomains, which also may function as exit sites for different virus species (see Table **1**). This association was successfully used to achieve incorporation of the GPI-anchored protein into the lipid envelope of VVs or VLPs [2,38,39], predominantly derived from retroviral particles. Co-transfection of plasmid vectors carrying genes for the production of retroviral vectors with constructs expressing the GPI-anchored proteins or super-transfection of pre-existing virus producing cell lines leads to the formation of viral particles decorated with GPI-anchored molecules on their envelopes (Fig. **1**, left). These particles acquire novel properties as a consequence of the incorporation of the GPI-anchored protein, *e.g.* super-transfection of the murine retroviral producer cell line PALSG/S with the human GPI-anchored protein CD59, yields viral particles that are resistant to the activity of complement in human serum [38]. These results suggested for the first time that incorporation of recombinantly expressed GPI-anchored proteins into the envelopes of VVs is possible and that these modifications can be useful for gene therapy approaches.

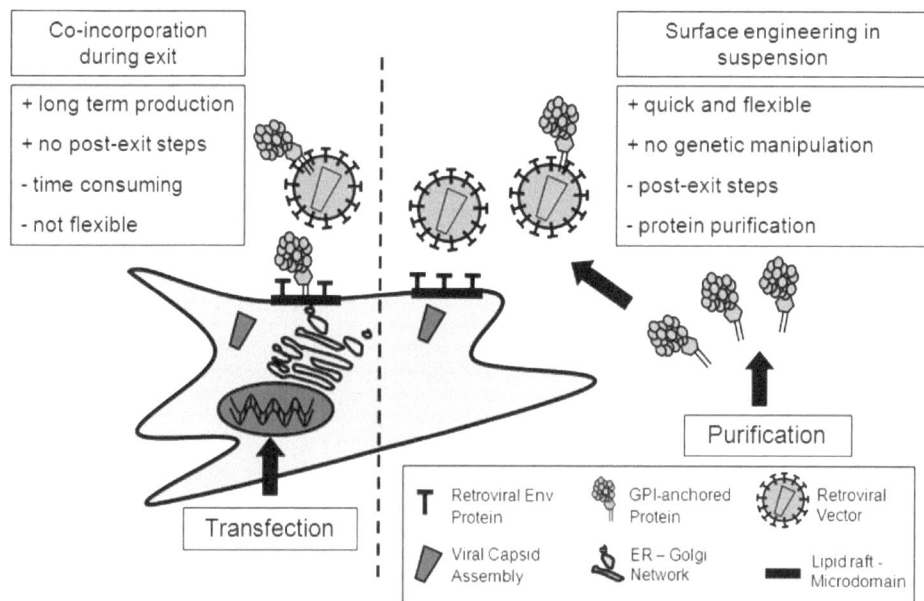

Figure 1: Modification of enveloped viruses, viral vectors and virus-like particles with glycosylphos-phatidylinositol-anchored proteins. Two different pathways are employed to engineer the surface of enveloped viral vectors. Transfection of virus producing cells with

constructs encoding for GPI-anchored leads to co-incorporation during exit from the cells. Alternatively, purified GPI-anchored proteins incubated with enveloped viruses, VVs and VLPs will insert into the virus envelope.

In two more recent studies, co-transfection approaches successfully generated VLPs displaying GPI-anchored molecules from mammalian2 or insect cells [39]. In both cases, recombinant GPI-anchored proteins were generated from different cytokine species i.e. interleukin-2 (IL-2) and granuolocyte-macrophage colony stimulating factor (GM-CSF) and tested for their functional properties. Kueng and co-workers demonstrated that the GPI-anchored cytokines can elicit cellular responses such as differentiation and proliferation with similar efficiency as their soluble counterparts; and Skountzou and co-workers showed that GPI-anchored cytokines engineered onto VLPs based on simian immunodeficiency virus (SIV) can be used to enhance immunogenicity of the VLPs in immunisation studies. Both approaches resulted in a modulation of immune responses provoked by the displayed GPI-anchored cytokine [2,39]. The major advantage of the transfection/co-localisation approach is that stable transfection of (retroviral) producer cell lines co- or super-transfected with endogenous or recombinant GPI-anchored proteins can provide a long-term, reliable source of modified viral particles with high reproducibility. Additionally, no post-exit steps that may reduce infectivity of VVs are required.

Modification of Viral Particles in Suspension

Transfection of producer cell lines is not the only possibility to use GPI-anchored proteins for modification of VVs. A process known as cellular "painting" [33,36] describes that GPI-anchored proteins, when extracted and purified from cells are re-inserted after incubation with eukaryotic cells and localisation is once again conferred to the plasma membrane.

This direct protein engineering or "painting" of enveloped viral particles occurs similar to cell painting when concentrated virus suspensions are incubated with purified GPI-anchored proteins (Fig. **1**, right). This was first described for the GPI-linked model protein CD59his which associates to viral vectors based on MLV and HIV-135, as well as feline herpesvirus (data unpub-lished). The association is specific and painted virus particles remain infectious after insertion of the GPI-linked protein, albeit at reduced efficiencies caused predominantly by the duration of the painting process, rather than the actual incorporation of GPI-anchored molecules into the viral envelope [35]. Estimates of the number of GPI-anchored proteins painted onto retroviral particles were in the range of the numbers observed for Env molecules per virion [35] and is thus similar to that achieved after incorporation of hybrid proteins produced in co-transfection experiments [39]. The main advantage of this approach is its flexibility. Different GPI-anchored proteins can be attached to a range of enveloped viral particles without repeated genetic manipulations of the virus-producing cells. This also means a considerable gain of time, compared to transfection-based methods. Additionally, the amounts of protein deposited at the viral surface may be controllable and only a limited amount of information about the genetic requirements of the virus is necessary for modification.

****

PROTOCOL 5

GPI Painting of Viruses, VVs and VLPs:

1. *Prepare concentrated virus stock from desired producer cell lines and purify desired GPI-anchored protein e.g. by affinity chromatography.*

2. *Incubate concentrated virus stocks with GPI-anchored proteins at desired concentration (5-50 µg/ml test range suggested for first trial) for more than 3 hr at 37°C under constant shaking.*

3. *Remove not associated GPI-anchored proteins from virus sample e.g. by ultracentrifugation or ultra-filtration protocols.*

4. *Analyse virus for presence of virus particles and GPI-anchored protein prior to application.*

****

Potential Applications

GPI-anchored proteins are useful for the modification of enveloped viral vectors using either of two profoundly different strategies: (i) transfection of viral producer cell lines and (ii) direct protein engineering of viral particles by painting (Fig. **1**). Viral painting may be the method of choice for modification of enveloped viral particles in all situations where a degree of flexibility is favourable, *e.g.* in response to genetic heterogeneity in gene therapy approaches or in response to high antigen variability for vaccination, and genetic modification of virus producing cell lines is difficult *e.g.* when applying toxic proteins or when handling genetically or biochemically poorly defined virus species are the targets of modification. Three main fields of interest for application are labelling of viral particles *e.g.* for capture or diagnosis, modulation of virus-host interactions *e.g.* immune-modulation for vaccine development and modification of virus binding properties *e.g.* for targeted delivery of VVs (Fig. **2**).

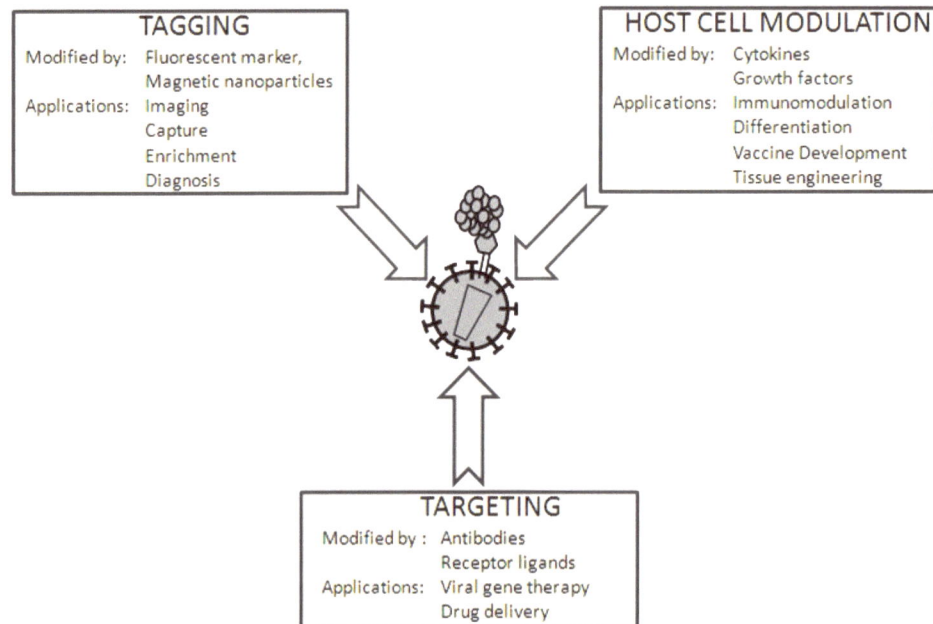

Figure 2: Possible applications for modifying viruses, VVs and VLPs with GPI-anchored proteins. Modification of viruses, VVs and VLPs using GPI-anchored proteins is useful for tagging of viruses, *e.g.* by using fluorescence markers for imaging or capture purposes. Modified viral surfaces can stimulate or inhibit host cell function, i.e. immune functions or induce differentiation. Modification also can re-direct binding specificities in order to target VVs or VLPs.

Virus Tagging

In its simplest application, GPI-anchored marker proteins may be used to identify and track viruses, VVs or VLPs. So far, viral particles are mainly labelled by incorporation of fusion proteins of viral structural proteins and (fluorescent) marker proteins [40,41]. However, these changes can also affect the biological behaviour of the particles. The use of GPI-anchored marker proteins provides an alternative, as the modification pathway does not interfere with viral structural components, i.e. the envelope surface protein required for target cell binding and infection. Furthermore, deposition of marker molecules on viral particles may be helpful for capture or enrichment prior to diagnostic procedures i.e. viral particles modified with 6 x his-tagged proteins can be enriched by using established immobilized metal ion affinity chromatography (IMAC) [42,43] or magnetic purification techniques [44]. This is especially interesting in cases where viruses tolerate concentration by centrifugation badly or where little is known about the virus, *e.g.* in the case of emergent viral diseases.

Modulation of Host Cell Functions

Interaction of virus particles with host molecules are often mediated by molecules located in the envelope. These interactions most importantly include immunological reactions. Modifying the envelope with immunologically competent molecules i.e. cytokines or growth factors allows for the manipulation of surrounding immune responses, thereby reducing possible side effects caused by systemic delivery of soluble agents [2]. This also includes protection of VVs from unwanted immune reactions such as complement activity [38]. Immunoprotection can help to ensure efficient delivery to target cells by eliminating premature inactivation of vectors in gene therapy approaches. Also presentation of antigen on the surface of VLPs is possible via the use of GPI-anchored molecules. These aspects can be of special importance in the development of vaccines or specific adjuvants enhancing vaccine efficacy. A further aspect of modulation of host cell function is inducing differentiation *e.g.* for tissue engineering purposes. Proof of principle was shown exemplary through differentiation of monocytes to dendritic cells [2].

Cellular Targeting

Viral glycoproteins located in the envelope function as recognition and entry devices to allow access to the target cells. Their specificity determines the host range or infection specificity of different virus species. Modification of these properties is crucial to achieve infection or trans-duction targeting – one of the most important goals of viral gene therapy. However, the direct manipulation of envelope glycoproteins i.e. by the generation of fusion proteins to redirect viral binding often leads to severely reduced infection efficiencies [45,46]. Deposition of specific binding factors by using GPI-anchored proteins may help to circumvent problems described previously, especially in cases where binding and entry of viral particles are mediated by independent factors [47,48]. Using viral painting technology under these circumstances has the advantage of flexibility. The same basic viral particle can be modified with a range of binding properties to suit the applications' specific needs. Antibody molecules, for example in form of single-chain antibody molecules, engineered to contain a GPI anchor, can provide a vast range of binding specificities. The same principle is applicable in drug delivery; however VVs and VLPs are less suitable for these approaches.

CELL SURFACE PAINTING WITH GPI-ANCHORED PROTEINS

Background

Cell surface engineering is a challenging but highly demanded task for biotechnological and medical applications. "Painting" a cell surface with a protein of choice is an attractive and powerful technology that can be achieved by decorating the plasma membrane with GPI-anchored proteins [33,49]. GPI-anchoring has the unique property that proteins with such a glypation moiety can stably incorporate into phospholipid bilayers of artificial planar membranes or cell membranes (Fig. **3**). This unusual ability of GPI-anchored proteins to re-insert into cell membranes was recognized in the framework of investigating functional aspects of the complement decay accelerating factor (DAF, CD55) to protect tissue from complement attacks. In a hallmark study, Medof and colleagues observed that the DAF protein preferentially re-incorporated into the surface of erythrocytes if the cells were incubated with purified membrane fractions of human red blood cells [50]. In fact, they described that purified DAF must have been re-inserted into the cell surface membrane, as DAF association with the erythrocyte's membrane could not be diminished by extensive washing of the cells even in the presence of high salt concentrations. However, the authors could extract membrane incorporated DAF with detergents. Moreover, they attributed some integral membrane properties to re-inserted DAF due to its functionality. This study is indeed remarkable as at that time the GPI-anchoring mechanism of DAF was not yet recognized, but followed three years later [51]. By elaborating on the incorporation phenomena of DAF, Medof and co-workers described that high- and low-density serum lipoproteins were able to interfere with the re-insertion [50]. In addition, membrane insertion of DAF occurred only at temperatures between 20 and 37°C, where a certain lateral mobility of lipids within the membrane is guaranteed.

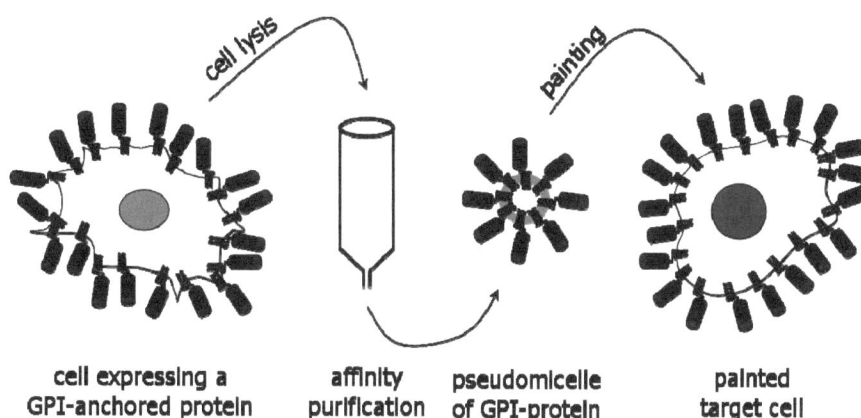

Figure 3: The principle of cell surface painting. Cells expressing an endogenous or recombinant GPI-anchored protein are lysed to generate a total protein extract which is used for affinity purification of the GPI-anchored protein. This can then be exogenously added to decorate the surface of a different cell or cell type.

State-of-the-Art

A pioneering study by Tykocinski and Medof then demonstrated that fusing the GPI anchor addition signal sequence (GSS) of DAF to the carboxyl terminus of the lymphocyte receptor CD8 permitted the expression of the fusion protein as GPI-anchored protein at the surface of transfected cells [37]. This study proved first evidences that basically any protein can be expressed as GPI-anchored protein. Both natural and artificially generated GPI-anchored proteins can be purified to homogeneity and re-inserted into the outer leaflet of virtually any target cell of choice. In a review article, Medof and co-workers nicely worked out major advantages of this technology of cell surface engineering [33]: any proteins can be generated as GPI-anchored protein. GPI-anchored proteins can be inserted into the plasma membrane even of cells that are difficult to transfect or to transduce. Membrane insertion of GPI-anchored proteins is fast, efficient, and remarkably stable and does not require previous cell culturing. Multiple GPI-anchored proteins can be painted concurrently or sequentially on the same cell. The number of painted molecules can be easily varied depending on the incubation time and the protein concentration used. Finally and most importantly, cell surface painted GPI-anchored proteins retain their natural functions [33]. Two publications describe in depth methodological aspects of how to design, express, purify and functionally re-insert GPI-anchored proteins [34,52]. One peculiar property of GPI-anchored proteins is to participate in so called membrane microdomains or lipid rafts [53,54]. Membrane compartmentalization plays a key role in signal transduction events, as it dictates aggregation and segregation of proteins at the membrane. Today, still many aspects of the role of such membrane microdomains are not fully understood. Due to their preferential localization in membrane microdomains, GPI-anchored GFP molecules are used as fluorescent markers to investigate membrane microdomains and membrane microdomain-associated proteins [36,55-57]. However, it was noted that, most probably due to different lipid compositions of the GPI-anchor, particularly on existence or absence of a palmitoylation at the inositol ring, different GPI-anchors differentially localize in membrane microdomains [36,58]. Noteworthy, the most stringent membrane microdomain localization was attributed to DAF, or GFP-GPI(DAF) [36]. Moreover, transfected and painted GPI-anchored proteins (re-)localized identically efficient in membrane microdomains [36,59]. Interestingly, mixing carrier lipids that are predominantly found in membrane microdomains, such as cholesterol, sphingomyelin or dipalmitoyl-phosphatidylethanolamine [60], with purified recombinant GPI-anchored proteins increased the efficiency of cell surface painting, whereas non-raft lipids had no effect [36].

****

PROTOCOL 6

Cell surface painting

1. *Wash target cells with pre-warmed PBS or serum-free medium by spinning cells down in a 15 ml tube. Certain GPI-anchored proteins have the tendency to stick to plastic, which may be avoided by using siliconized tubes.*
2. *Re-suspend cells in serum-free pre-warmed medium to a density of 10^6 to 10^7 cells per ml. Cells sensitive to serum-starvation may be painted in serum-containing medium with higher concentrations of GPI-anchored proteins.*
3. *Purified GPI-anchored proteins are added to a final concentration of 0.5 to 5 μg/ml.*
4. *Cells are incubated with purified GPI-anchored proteins at 37°C/5%CO$_2$ for 30 minutes to several hours. For many cell types cell surface painting reaches a plateau of insertion after 90 minutes of incubation. Cell sedimentation is prevented by occasional gentle mixing permitting efficient painting.*
5. *Extensively wash painted cells and verify painting efficiency by flow cytometry.*

****

Biomedical Applications

The technique of cell surface painting is particularly interesting for biomedical applications. For example, erythrocytes that were painted with the adhesion molecule LFA-1 were then able to interact with target cells expressing its counter-receptor CD2 [61]. Decorating cells with an artificial high affinity ligand for the integrin αvβ3, termed KISS31, gained adhesive properties to interact with immobilized αvβ3 integrin *in vitro* [62]. As the integrin αvβ3 is up-regulated in tumour-associated blood vessels [63] targeting tumour-specific cytotoxic T cells to αvβ3 integrin-expressing tumours was proposed as strategy to fight against cancer. In fact, primary T cells painted with GPI-anchored KISS31 adhered to αvβ3 integrin *in vitro* and, if injected into tumour-bearing mice, homed to αvβ3 integrin-positive tumours *in vivo* [62]. Moreover, painting of target cells with peptide-loaded GPI-anchored MHC I complexes were efficiently recognized and killed by antigen-specific cytotoxic T cells *in vitro* [64]. Cell surface painting of GPI-anchored co-stimulatory molecules, such as CD80 and CD86, was shown to enhance cell proliferation [65]. Injection of murine tumour cell membranes, that were previously painted *in vitro* with GPI-anchored CD80, enhanced the expansion of tumour-specific cytotoxic T cells *in vivo* [66]. Furthermore, immunization with CD80-GPI-painted tumour cell membranes protected mice from a parental tumour challenge [66]. Further experiments are needed to exploit the painting strategy, particularly by re-inserting multiple GPI-anchored proteins in cells, for targeting tumours and other diseases *in vivo*. Cell painting may also be an alternative to *ex vivo* gene therapy approaches, i.e. instead of transduction of therapeutic target cells, for example with retroviral vectors, the cells can be decorated with GPI-anchored proteins.

Beside above described efforts to enhance immunity, there is another interesting therapeutic potential of GPI-anchored proteins: Patients suffering from paroxysmal nocturnal hemoglobinuria bear a somatic mutation in the PIGA gene located on the X chromosome [67]. The enzyme PIGA is pivotal for the synthesis of the GPI moiety. In fact, patients with paroxysmal nocturnal hemoglobinuria show aberrant expression of the GPI-anchored protein DAF [68]. Consequently, erythrocytes from these patients are destroyed by complement-mediated lysis due to the lack of the complement-inhibitory function of DAF. Interestingly, cell-to-cell transfer of DAF was observed to occur *in vivo* in a transgenic mouse model [68]. This observation let to a clinical trial, where six patients suffering from paroxysmal nocturnal hemoglobinuria were transfused with erythrocytes derived from healthy donors [69]. At day one, three and seven after transfusion, DAF surface expression was detectable on recipient erythrocytes and granulocytes in all six patients [69], indicating that DAF derived from healthy cells were transferred simultaneously to deficient cells *in vivo*. This study suggests that simple transfusion and *in vivo* transfer of GPI-anchored DAF may have tremendous therapeutic potential to treat paroxysmal nocturnal hemoglobinuria patients.

SURFACE ENGINEERING OF MICRO- AND NANOPARTICLES WITH GPI-ANCHORED PROTEINS

Background

From the surface modification of cells, followed by enveloped viruses and viral vectors, the natural progression for GPI painting lies with micro- and nanoparticles, most obviously for those that are cell membrane derived or biomembrane-like. Giant unilamellar vesicles (GUVs) are in the ten to hundred micrometer range and can be produced by mixing diverse phospholipids and membrane components such as cholesterol and phosphatidylcholines under static electric current. They are widely used in membrane research due to their size and ease of handling for studying lipid phase separation and the appearance and behaviour of lipid raft domains [70]. Liposomes are also produced by mixing phospholipids in aqueous solution and, although varying in size, they are generally 2-3 dimensions smaller than GUVs. Liposomes have often been proposed as therapeutic agents as they are seemingly ideal in many ways [71]. They have an aqueous centre which can be used to carry drugs or therapeutic molecules (membrane bound if required), they can fuse with cells to deliver the therapeutic agent directly into or onto the target cell, they are non-toxic to biological systems and there are many modern biochemical possibilities for modifying their exterior surface to make them biocompatible, via PEGylation for example [72], or to target them (Fig. **4**). Other closely related, manmade particles of this dimension are micelles (the same as liposomes but with no aqueous centre) and nanoemulsions or various kinds of polymeric nanoparticles [71]. All such nanocarriers can have lipid bilayer or lipid bilayer-like properties and may have the potential for GPI modification.

Liposome for Drug Delivery

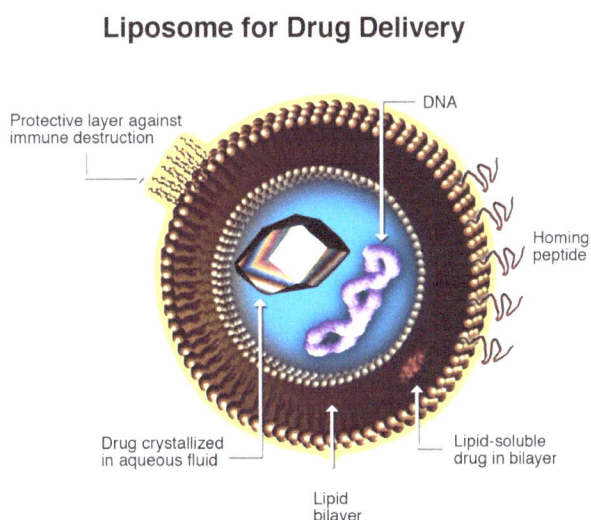

Figure 4: The concept of a multifunctional, liposome based nanocarrier as depicted by Kosi Gramatikoff [73]. Soluble drugs or therapeutic biomolecules (*e.g.* DNA, RNA, peptides, proteins, antibodies) can be incorporated into the aqueous centre and non-soluble drugs or molecules into the lipid bilayer. Immuno-protective and targeting agents can be incorporated chemically or potentially by means of GPI painting onto the surface.

Lipid bilayer membrane particles derived from mammalian cells, commonly known as membrane vesicles, typically lie in a size range of ten to low hundreds of nanometers. Exosomes, usually described as being in the 40 – 100 nm range, are generated by membrane invagination into multivesicular bodies within the cytoplasm and then released from the cell by exocytosis. Once outside, exosomes can also be referred to as membrane vesicles or micro-vesicles although such particles can also be formed by proteolytic cleavage of surface GPI-anchored proteins, apoptosis and mechanical stress to the cell membrane (and are as such a common component of laboratory cell culture systems) [74]. Generally, such vesicles have been widely studied in the field of cellular and molecular biology since the early 1960's but due to their composition, size and density it is worth to mention that they have often been confused with viruses and virus-like particles and *vice-versa*. Exosomes have been widely studied in their role for intercellular communication but, by containing microbial components, are also strongly linked with the cell-to-cell spread of infectious agents [75]. As they can be found in many easily accessible bodily fluids, such as blood and saliva, there is a great interest recently in their use as potential disease markers for diagnostics [76] and it is especially in this respect that a connection with GPI painting can be made.

Inorganic microparticles can also be surface engineered with GPI-anchored proteins, but in contrast to cells and viruses the GPI-anchor is not used, instead other chemistries available on the protein part of the GPI molecule, such as amino acid-based tags or individual amino acid modifications (allowing for example biotinylation), are be used for attachment. Although this is not painting in itself, this leaves the GPI anchor free for membrane attachment thereafter, adding an extra dimension to the painting system. Hence, correctly engineered GPI-anchored proteins may not only be painted onto to a membrane as a biofunctional unit but at the same time used as a linker to inorganic particles such as magnetic nanoparticles or quantum dots for many potential applications, i.e. biomembrane surfaces can be simultaneously functionalised not only with multiple active GPI-anchored proteins but also with diverse nanomaterials.

State of the Art

Synthetic Membrane Vesicles

Aside from the use of GUVs and liposomes as tools for the basic understanding of GPI-anchored protein biology, the use of GPI painting for modification of such particles can still be considered as an emerging area and published work remains sparse. However, there is a lot of current activity and results are starting to emerge. Bumgarner and co-workers have invented and characterised a method to engineer the surface of biodegradable microparticles with GPI-anchored proteins [77]. This stemmed from many earlier years of work from the same group of Selvaraj on cell painting and cancer vaccine development showing that most likely any protein can be engineered to have a GPI-anchor and stays functional as such [34,66,78-80]. In this recent study, albumin based microparticles (between 4.3 and 4.5 μm depending on the production technique), were painted with the immune regulatory proteins GPI-hB7-1 (or CD80) and GPI-mICAM-1 (or CD54) (Fig. **5**). The association between GPI-anchored protein and microparticle was characterised and optimised in terms of concentration, temperature and kinetics. As with cells and viruses, the association was shown to be reversible and dependent on the presence of an intact GPI-anchor. Although there is clearly an activity window, storage conditions over a week at 4°C were considered to be reasonable. Interestingly, both molecules could be simultaneously painted onto the same particles without any negative influence being seen, however, functionality of the molecules so far has only been demonstrated by means of antibody binding, so their true effectiveness in a medical setting still needs to be shown [77].

****

PROTOCOL 7

Microparticle surface modification with GPI-anchored proteins:

1. *Suspend 1 mg of microparticles (this protocol is optimized for albumin microparticles) in 150 μl of PBS (pH 7.4) in a 1.5 ml microcentrifuge tube and incubate with 20 μg/ml of the GPI-anchored protein of interest for 20 mins at room temperature (the microcentrifuge tubes should be lightly tapped every 5 mins to insure good mixing).*

2. *Fill up the microcentrifuge tube with wash buffer (PBS/5 mM EDTA/1% FCS), centrifuge briefly to pellet the microparticles bearing the GPI-anchored protein and remove the supernatant by means of suction.*

3. *Repeat step 2 with 200 μl of wash buffer and then plate in triplicate 50 μl of microparticles in wash buffer into the wells of a 96-well V-bottom plate.*

4. *Add 50 μl of antibody directed against the GPI-anchored protein of interest to the wells (20 μg/ml) and shake the plate for 30 mins at 4°C on a plate shaker.*

5. *Wash the microparticles three times using wash buffer by filling the wells, centrifuging the plate to pellet the microparticles and removing the supernatant by suction.*

6. *After the 3rd wash, resuspend the pelleted microparticles in 50 µl of a suitable HRP-conjugated secondary antibody and incubate the plate for 30 mins at 4°C on a plate shaker.*

7. *Wash the microparticles two times with wash buffer by filling the wells, centrifuge the plate to pellet the microparticles and remove the supernatant by suction.*

8. *Resuspend the microparticles in PBS (pH 7.4) for the final wash, centrifuge the plate and resuspend the pelleted microparticles in 100 µl of peroxidase substrate TMB and observe for color development.*

9. *Stop the color development with 2N H_2SO_4 and centrifuge the V-bottomed plate at 1000 x g to pellet the microparticles.*

10. *Transfer 100 µl of the supernatant into a 96-well flat-bottomed ELISA plate and assay the color development at 415 nm using a microplate reader to verify binding of the GPI-anchored protein to the surface of the microparticle.*

11. *As a control for the reaction, incubate 1 mg of microparticles with 150 µl of buffer alone without the presence of the GPI-anchored protein.*

****

Link to Nanotechnology

The last 5 to 10 years has seen dramatic increase of inter-disciplinary work to combine sub-micron materials such as diverse inorganic nanoparticles or carbon nanotubes with biological components such as nucleic acids, peptides, proteins or antibodies leading to the research area now known as bionanotechnology and there is much evidence to suggest that inventions from this area will play a big role in the future of medicine [82]. GPI-surface functionalisation is able to play a significant role here. The GPI-GFP-his and CD59his proteins used for painting of viral vectors were genetically modified to contain a 10 x histidine tag and this was specifically engineered into hydrophilic regions of the proteins which are regions protruding outwards from membrane binding region (35 and unpublished data). Histidine tags are able to bind metals such as gold, cobalt and nickel, whereby nickel nitrilo-tri-acetic acid (Ni-NTA) is the strongest due to its ion availability [83]. This was done not only to optimise the purification

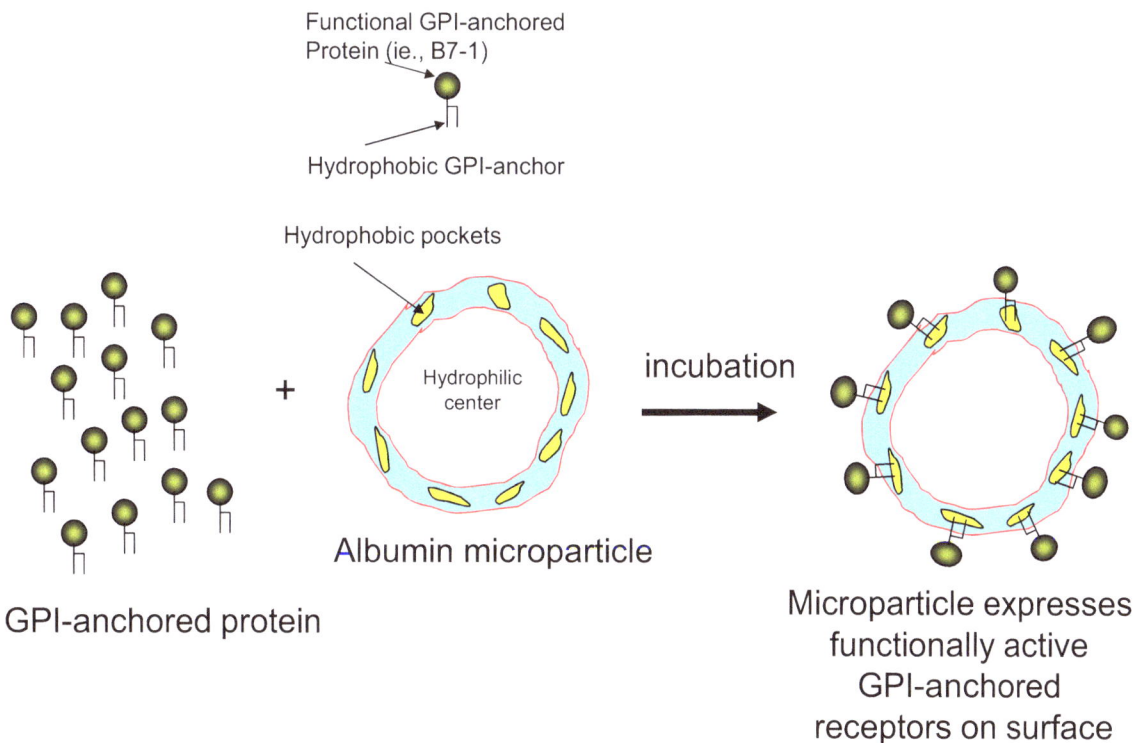

Figure 5: Schematic representation of the protein transfer process onto albumin microparticles [81].

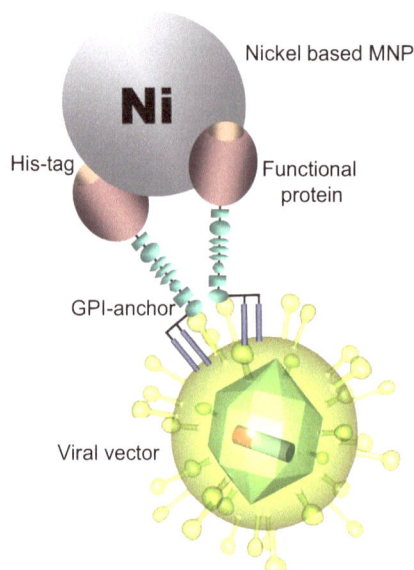

Figure 6: Linking organic and inorganic particles with GPI-anchored proteins. Proteins can be engineered to contain a GPI-anchor as well as a tag (*e.g.* his-tag) and remain functional. As such they can be used to link any membrane based entity such as a cell, virus, liposome etc (an enveloped retroviral vector is depicted here) to a nanomaterial designed to have a particular function for targeting, imaging etc (here a nickel based magnetic nanoparticle is shown).

procedure and alleviate the necessity for more expensive and timely antibody related methods, but also to allowspecific and orientation controllable linkage for the purified GPI-anchored proteins to virus-sized, nickel-based magnetic nanoparticles (Ni-MNPs). GPI-GFP-his and CD59his have been shown to bind several of such custom made Ni-MNPs, whereby iron core, Ni-NTA coated MNPs [84] were shown to have the highest binding efficiency and are also expected to be more biocompatible than pure nickel particles (unpublished data). Functionalised particles of this nature can then be painted onto the surface of viral vectors, cells, liposomes or membrane vesicles (Fig. **6**).

Potential Applications

GPI-Anchored Protein Production

When considering the use of GPI-anchored proteins for exogenous surface engineering, whether it be of cells, viruses or other particles, an important aspect is the production of the GPI-anchored protein itself. As bacteria do not make GPI-anchored proteins, expression systems are all based on eukaryotes. Currently, purification from mammalian cells is achieved by engineering in tags such as the histidine tag, by ion exchange chromatography or by using antibody related chro-matographical methods such as immuno-affinity or immuno-precipitation [35,36,62,77,85]. In bio-logical terms, the process of GPI anchoring was shown to be critically important by the fact that its abrogation leads to embryonic lethality [86], however, only around 1% of proteins are GPI-anchored in mammals. This is likely due, at least in part, to it being a costly and complicated process, requiring more than twenty gene products [87]. Therefore, yields of recombinant GPI-anchored proteins from mammalian expression systems are rather poor and purification can be technically tricky due to insolubility issues, i,e. the tendency to form micelles under aqueous conditions and in the absence of detergents. Despite their costs, there are some non-toxic detergents available for this (which are necessary for biological compatibility thereafter), but for future applications or off-the-shelf products this and the other obstacles described above will need to be overcome. One option is to optimise the production in mammalian cells. Although there is some evidence from work in insect cells that up-regulation of genetic factors within the GPI biosynthetic pathway may be an option [88], due to the multiplicity of factors involved, this approach may be limited and the most likely resort will be to up-scale the process. This is likely to be impractical and costly. One possibility is the completely synthetic production of GPI mimics or GPI-like molecules. Chemical biologists have already made significant progress here and this topic is addressed in Chapter 5. If one is to remain with biological approaches, expression in yeast systems may be interesting because of the far easier possibility of process up-scaling. For cell-surface display purposes, it was seen that recombinant GPI-anchored proteins can be efficiently expressed in yeast [89,90]. Although there are some differences between mammalian and yeast GPIs, such as variations in the number of lipid prongs, recently it was found that human CD59his could be expressed in yeast (unpublished data), showing their potential as a GPI expression system per se.

Synthetic Membrane Vesicles

The interest for GPI surface modification of GUVs is likely to remain in the research area but for liposomes and other related nanocarriers, the key application clearly lies with the delivery of drugs or therapeutic molecules [71]. In this respect GPI modification may allow a simpler and quicker means (compared to chemical linkage which may also affect the functional activity of the molecule) to modify the surface of liposomes with proteins to target specific cells (*e.g.* receptor ligands), give immune-camouflage until they have reached their target (*e.g.* compliment regulatory molecules such as CD59) or link the liposome with nanomaterials for targeting and concentration at the therapeutic site (*e.g.* magnetic nanoparticles). Evidence already suggests that painting of albumin bioparticles has great potential for immune stimulation for vaccine approaches (*e.g.* adjuvant activity) and in turn enhance capacity as a targeted antigen or drug delivery device [81].

Bio-derived Membrane Microparticles

Membrane vesicles or micro-vesicles (MV) are interesting for GPI painting applications in a number of areas. MV can always be found in laboratory cell culture supernatants and several groups have activities in developing conditioned mediums as therapeutic agents. For example, evidence suggests that large complexes (100 – 220 nm) found in conditioned media from human mesenchymal stem cells are the agents responsible for significant cardioprotective effects in mouse and pig myocardial infarction models [91]. In the same way as discussed for liposomes above, MV may be functionalised with GPI-anchored molecules for targeting or immune-protection in such approaches. Many stem cell culture systems rely on the use of conditioned medium from so called feeder cells [92]. As it is highly likely that MV are an important component of such mediums, this opens the possibility for characterisation or enhancement of such mediums through GPI painting.

MV, in the form of exosomes, can be readily found in blood, urine, saliva, breast milk and a range of other bodily fluids and there is an ever increasing interest in their use as blood borne diagnostic markers for many diseases, including cancer [74-76,93-97]. Painting may have a diagnostics application in this respect as clinical samples could be painted with GPI tagged magnetic nanoparticles for quick and easy isolation and detection of specific MV disease markers. Similarly to what has been proposed for viruses, it may also be possible to introduce painted vesicles *in vivo* in order to track and image them for the basic understanding of the roles they play in the body or for monitoring disease progression.

REFERENCES

[1]   Roy, P. & Noad, R. Virus-like particles as a vaccine delivery system: myths and facts. Hum Vaccin 4, 5-12 (2008).
[2]   Kueng, H. J. et al. General strategy for decoration of enveloped viruses with functionally active lipid-modified cytokines. J Virol 81, 8666-76 (2007).
[3]   Barman, S. & Nayak, D. P. Analysis of the transmembrane domain of influenza virus neuraminidase, a type II transmembrane glycoprotein, for apical sorting and raft association. J Virol 74, 6538-45 (2000).
[4]   Scheiffele, P. et al. Interaction of influenza virus haemagglutinin with sphingolipid-cholesterol membrane domains via its transmembrane domain. EMBO J 16, 5501-8 (1997).
[5]   Zhang, J. et al. The cytoplasmic tails of the influenza virus spike glycoproteins are required for normal genome packaging. Virology 269, 325-34 (2000).
[6]   Manie, S. N. et al. Measles virus structural components are enriched into lipid raft microdomains: a potential cellular location for virus assembly. J Virol 74, 305-11 (2000).
[7]   Vincent, S. et al. Measles virus assembly within membrane rafts. J Virol 74, 9911-5 (2000).
[8]   Marty, A. et al. Association of matrix protein of respiratory syncytial virus with the host cell membrane of infected cells. Arch Virol 149, 199-210 (2004).

[9]   McDonald, T. P. et al. Evidence that the respiratory syncytial virus polymerase complex associates with lipid rafts in virus-infected cells: a proteomic analysis. Virology 330, 147-57 (2004).

[10]   Brown, G. et al. Respiratory syncytial virus assembly occurs in GM1-rich regions of the host-cell membrane and alters the cellular distribution of tyrosine phosphorylated caveolin-1. J Gen Virol 83, 1841-50 (2002).

[11]   Ali, A. & Nayak, D. P. Assembly of Sendai virus: M protein interacts with F and HN proteins and with the cytoplasmic tail and transmembrane domain of F protein. Virology 276, 289-303 (2000).

[12]   Brown, E. L. & Lyles, D. S. Organization of the vesicular stomatitis virus glycoprotein into membrane microdomains occurs independently of intracellular viral components. J Virol 77, 3985-92 (2003).

[13]   Aman, M. J. et al. Molecular mechanisms of filovirus cellular trafficking. Microbes Infect 5, 639-49 (2003).

[14]   Bavari, S. et al. Lipid raft microdomains: a gateway for compartmentalized trafficking of Ebola and Marburg viruses. J Exp Med 195, 593-602 (2002).

[15]   Panchal, R. G. et al. In vivo oligomerization and raft localization of Ebola virus protein VP40 during vesicular budding. Proc Natl Acad Sci U.S.A. 100, 15936-41 (2003).

[16]   Nguyen, D. H. & Hildreth, J. E. Evidence for budding of human immunodeficiency virus type 1 selectively from glycolipid-enriched membrane lipid rafts. J Virol 74, 3264-72 (2000).

[17]   Ono, A. & Freed, E. O. Plasma membrane rafts play a critical role in HIV-1 assembly and release. Proc Natl Acad Sci U.S.A. 98, 13925-30 (2001).

[18]   Pickl, W. F. et al. Lipid rafts and pseudotyping. J Virol 75, 7175-83 (2001).

[19]   Gilbert, J., Dahl, J., Riney, C., You, J., Cui, C., Holmes, R., Lencer, W. & Benjamin, T. Ganglioside GD1a restores infectibility to mouse cells lacking functional receptors for polyomavirus. J Virol 79, 615-8 (2005).

[20]   Grassme, H. et al. Rhinoviruses infect human epithelial cells via ceramide-enriched membrane platforms. J Biol Chem 280, 26256-62 (2005).

[21]   Colin, M. et al. Efficient species C HAdV infectivity in plasmocytic cell lines using a clathrin-independent lipid raft/caveola endocytic route. Mol Ther 11, 224-36 (2005).

[22]   Delmas, O. et al. Spike protein VP4 assembly with maturing rotavirus requires a postendoplasmic reticulum event in polarized caco-2 cells. J Virol 78, 10987-94 (2004).

[23]   Hancock, J. F. Lipid rafts: contentious only from simplistic standpoints. *Nat Rev Mol* Cell Biol 7, 456-62 (2006).

[24]   Hanzal-Bayer, M. F. & Hancock, J. F. Lipid rafts and membrane traffic. FEBS Lett 581, 2098-104 (2007).

[25]   Michel, V. & Bakovic, M. Lipid rafts in health and disease. Biol Cell 99, 129-40 (2007).

[26]   Shaw, A. S. Lipid rafts: now you see them, now you don't. Nat Immunol 7, 1139-42 (2006).

[27]   Suzuki, T. & Suzuki, Y. Virus infection and lipid rafts. Biol Pharm Bull 29, 1538-41 (2006).

[28]   Chazal, N. & Gerlier, D. Virus entry, assembly, budding, and membrane rafts. Microbiol Mol Biol Rev 67, 226-37, table of contents (2003).

[29]   Metzner, C. et al. Rafts, anchors and viruses--a role for glycosylphosphatidylinositol anchored proteins in the modification of enveloped viruses and viral vectors. Virology 382, 125-31 (2008).

[30]   Saifuddin, M. et al. Human immunodeficiency virus type 1 incorporates both glycosyl phosphatidylinositol-anchored CD55 and CD59 and integral membrane CD46 at levels that protect from complement-mediated destruction. J Gen Virol 78 (Pt 8), 1907-11 (1997).

[31]   Banki, Z. et al. HIV and human complement: inefficient virolysis and effective adherence. Immunol Lett 97, 209-14 (2005).

[32]   Stoiber, H. et al. Complement-HIV interactions during all steps of viral pathogenesis. Vaccine 26, 3046-54 (2008).

[33]   Medof, M. E. et al. Cell-surface engineering with GPI-anchored proteins. FASEB J 10, 574-86 (1996).

[34]   Nagarajan, S. et al. Purification and optimization of functional reconstitution on the surface of leukemic cell lines of GPI-anchored Fc gamma receptor III. J Immunol Methods 184, 241-51 (1995).

[35]   Metzner, C. et al. Association of glycosylphosphatidylinositol-anchored protein with retroviral particles. FASEB J 22, 2734-9 (2008).

[36]   Legler, D. F. et al. Differential insertion of GPI-anchored GFPs into lipid rafts of live cells. FASEB J 19, 73-5 (2005).

[37]   Tykocinski, M. L. et al. Glycolipid reanchoring of T-lymphocyte surface antigen CD8 using the 3' end sequence of decay-accelerating factor's mRNA. Proc Natl Acad Sci U.S.A. 85, 3555-9 (1988).

[38]   Breun, S. et al. Protection of MLV vector particles from human complement. Biochem Biophys Res Commun 264, 1-5 (1999).

[39] Skountzou, I. et al. Incorporation of glycosylphosphatidylinositol-anchored granulocyte- macrophage colony-stimulating factor or CD40 ligand enhances immunogenicity of chimeric simian immunodeficiency virus-like particles. J Virol 81, 1083-94 (2007).

[40] Lehmann, M. J. et al. Actin- and myosin-driven movement of viruses along filopodia precedes their entry into cells. *J* Cell Biol 170, 317-25 (2005).

[41] Lampe, M. et al. Double-labelled HIV-1 particles for study of virus-cell interaction. Virology 360, 92-104 (2007).

[42] Gaberc-Porekar, V. & Menart, V. Perspectives of immobilized-metal affinity chromatography. J Biochem Biophys Methods 49, 335-60 (2001).

[43] Magnusdottir, A. et al. Enabling IMAC purification of low abundance recombinant proteins from E. coli lysates. Nat Methods 6, 477-8 (2009).

[44] Franzreb, M. et al. Protein purification using magnetic adsorbent particles. Appl Microbiol Biotechnol 70, 505-16 (2006).

[45] Galanis, E. et al. Delivery systems intended for in vivo gene therapy of cancer: targeting and replication competent viral vectors. Crit Rev Oncol Hematol 38, 177-92 (2001).

[46] Zhao, T. M. et al. Infectivity of chimeric human T-cell leukemia virus type I molecular clones assessed by naked DNA inoculation. Proc Natl Acad Sci U.S.A. 93, 6653-8 (1996).

[47] Lin, A. H. et al. Receptor-specific targeting mediated by the coexpression of a targeted murine leukemia virus envelope protein and a binding-defective influenza hemagglutinin protein. Hum Gene Ther 12, 323-32 (2001).

[48] Yang, L. et al. Targeting lentiviral vectors to specific cell types in vivo. Proc Natl Acad Sci U.S.A. 103, 11479-84 (2006).

[49] Tykocinski, M. L. et al. New designs for cancer vaccine and artificial veto cells: an emerging palette of protein paints. Immunol Res 27, 565-74 (2003).

[50] Medof, M. E. et al. Inhibition of complement activation on the surface of cells after incorporation of decay-accelerating factor (DAF) into their membranes. J Exp Med 160, 1558-78 (1984).

[51] Caras, I. W. et al. Signal for attachment of a phospholipid membrane anchor in decay accelerating factor. Science 238, 1280-3 (1987).

[52] Fayen, J. D. et al. Glycerolphosphoinositide anchors for membrane-tethering proteins. Methods Enzymol 327, 351-68 (2000).

[53] Simons, K. & Toomre, D. Lipid rafts and signal transduction. Nat Rev Mol Cell Biol 1, 31-9 (2000).

[54] Anderson, R. G. & Jacobson, K. A role for lipid shells in targeting proteins to caveolae, rafts, and other lipid domains. Science 296, 1821-5 (2002).

[55] Nichols, B. J. et al. Rapid cycling of lipid raft markers between the cell surface and Golgi complex. J Cell Biol 153, 529-41 (2001).

[56] Hofman, E. G. et al. EGF induces coalescence of different lipid rafts. J Cell Sci 121, 2519-28 (2008).

[57] Choudhury, A. et al. Regulation of caveolar endocytosis by syntaxin 6-dependent delivery of membrane components to the cell surface. Nat Cell Biol 8, 317-28 (2006).

[58] Madore, N. et al. Functionally different GPI proteins are organized in different domains on the neuronal surface. EMBO J 18, 6917-26 (1999).

[59] Premkumar, D. R. et al. Properties of exogenously added GPI-anchored proteins following their incorporation into cells. J Cell Biochem 82, 234-45 (2001).

[60] Legler, D. F. et al. Selective inhibition of CTL activation by a dipalmitoyl-phospholipid that prevents the recruitment of signaling molecules to lipid rafts. FASEB J 15, 1601-3 (2001).

[61] Selvaraj, P. et al. Deficiency of lymphocyte function-associated antigen 3 (LFA-3) in paroxysmal nocturnal hemoglobinuria. Functional correlates and evidence for a phosphatidylinositol membrane anchor. J Exp Med 166, 1011-25 (1987).

[62] Legler, D. F. et al. The alpha v beta 3 integrin as a tumor homing ligand for lymphocytes. Eur J Immunol 34, 1608-16 (2004).

[63] Gasparini, G. et al. Vascular integrin alpha(v)beta3: a new prognostic indicator in breast cancer. Clin Cancer Res 4, 2625-34 (1998).

[64] Huang, J. H. et al. Protein transfer of preformed MHC-peptide complexes sensitizes target cells to T cell cytolysis. Immunity 1, 607-13 (1994).

[65] Brunschwig, E. B. et al. Protein transfer of glycosyl-phosphatidylinositol (GPI)-modified murine B7-1 and B7-2 costimulators. J Immunother 22, 390-400 (1999).

[66] McHugh, R. S. et al. Protein transfer of glycosyl-phosphatidylinositol-B7-1 into tumor cell membranes: a novel approach to tumor immunotherapy. Cancer Res 59, 2433-7 (1999).

[67] Parker, C. et al. Diagnosis and management of paroxysmal nocturnal hemoglobinuria. Blood 106, 3699-709 (2005).

[68] Carothers, D. J. et al. Synthesis of aberrant decay-accelerating factor proteins by affected paroxysmal nocturnal hemoglobinuria leukocytes. J Clin Invest 85, 47-54 (1990).

[69] Sloand, E. M. et al. Transfer of glycosylphosphatidylinositol-anchored proteins to deficient cells after erythrocyte transfusion in paroxysmal nocturnal hemoglobinuria. Blood 104, 3782-8 (2004).

[70] Wesolowska, O. et al. Giant unilamellar vesicles - a perfect tool to visualize phase separation and lipid rafts in model systems. Acta Biochim Pol 56, 33-9 (2009).

[71] Torchilin, V. P. Multifunctional nanocarriers. Adv Drug Deliv Rev 58, 1532-55 (2006).

[72] Harris, J. M. et al. Pegylation: a novel process for modifying pharmacokinetics. Clin Pharmacokinet 40, 539-51 (2001).

[73] Gramatikoff, K. "Liposome". I, the copyright holder of this work, hereby release it into the public domain. This applies worldwide. In case this is not legally possible, I grant any entity the right to use this work for any purpose, without any conditions, unless such conditions are required by law. Sourced from Wikipedia (1999).

[74] Pisitkun, T. et al. Discovery of urinary biomarkers. Mol Cell Proteomics 5, 1760-71 (2006).

[75] Schorey, J. S. & Bhatnagar, S. Exosome function: from tumor immunology to pathogen biology. Traffic 9, 871-81 (2008).

[76] Simpson, R. J. et al. Exosomes: proteomic insights and diagnostic potential. Expert Rev Proteomics 6, 267-83 (2009).

[77] Bumgarner, G. W. et al. Surface engineering of microparticles by novel protein transfer for targeted antigen/drug delivery. J Control Release 137, 90-7 (2009).

[78] McHugh, R. S. et al. Construction, purification, and functional incorporation on tumor cells of glycolipid-anchored human B7-1 (CD80). Proc Natl Acad Sci U.S.A. 92, 8059-63 (1995).

[79] Cimino, A. M. et al. Cancer vaccine development: protein transfer of membrane-anchored cytokines and immunostimulatory molecules. Immunol Res 29, 231-40 (2004).

[80] Selvaraj, P. et al. Custom designing therapeutic cancer vaccines: delivery of immunostimulatory molecule adjuvants by protein transfer. Hum Vaccin 4, 384-8 (2008).

[81] Reprinted from Journal of Controlled Release: Bumgarner, G. W., Shashidharamurthy, R., Nagarajan, S., D'Souza, M. J. & Selvaraj, P. Surface engineering of microparticles by novel protein transfer for targeted antigen/drug delivery. J Control Release **137**, 90-7 (2009). Copyright (2009), with permission from Elsevier.

[82] Gao, J. et al. Multifunctional Magnetic Nanoparticles: Design, Synthesis, and Biomedical Applications. Acc Chem Res (2009).

[83] Patel, J. D. et al. Preparation and Characterization of Nickel Nanoparticles for Binding to His-tag Proteins and Antigens. Pharm Res 24, 343-52 (2007).

[84] Lim, Y. T. et al. Immobilization of histidine-tagged proteins by magnetic nanoparticles encapsulated with nitrilotriacetic acid (NTA)-phospholipids micelle. Biochem Biophys Res Commun 344, 926-30 (2006).

[85] Brodsky, R. A. et al. Purified GPI-anchored CD4DAF as a receptor for HIV-mediated gene transfer. Hum Gene Ther 5, 1231-9 (1994).

[86] Nozaki, M. et al. Developmental abnormalities of glycosylphosphatidylinositol-anchor-deficient embryos revealed by Cre/loxP system. Lab Invest 79, 293-9 (1999).

[87] Orlean, P. & Menon, A. K. Thematic review series: lipid posttranslational modifications. GPI anchoring of protein in yeast and mammalian cells, or: how we learned to stop worrying and love glycophospholipids. J Lipid Res 48, 993-1011 (2007).

[88] Shams-Eldin, H. et al. An efficient method to express GPI-anchor proteins in insect cells. Biochem Biophys Res Commun 365, 657-63 (2008).

[89] Adachi, T. et al. Construction of an Aspergillus oryzae cell-surface display system using a putative GPI-anchored protein. Appl Microbiol Biotechnol 81, 711-9 (2008).

[90] Murai, T. et al. Construction of a starch-utilizing yeast by cell surface engineering. Appl Environ Microbiol 63, 1362-6 (1997).

[91] Timmers, L. et al. Reduction of myocardial infarct size by human mesenchymal stem cell conditioned medium. Stem Cell Res 1, 129-37 (2007).

[92] Choo, A. et al. Immortalized feeders for the scale-up of human embryonic stem cells in feeder and feeder-free conditions. J Biotechnol 122, 130-41 (2006).

[93] Pisitkun, T. et al. Identification and proteomic profiling of exosomes in human urine. Proc Natl Acad Sci U.S.A. 101, 13368-73 (2004).

[94] Horstman, L. L. et al. Cell-derived microparticles and exosomes in neuroinflammatory disorders. Int Rev Neurobiol 79, 227-68 (2007).

[95] Ichim, T. E. et al. Exosomes as a tumor immune escape mechanism: possible therapeutic implications. J Transl Med 6, 37 (2008).

[96] Mitchell, P. J. et al. Can urinary exosomes act as treatment response markers in prostate cancer? *J* Transl Med 7, 4 (2009).

[97] Rabinowits, G. et al. Exosomal microRNA: a diagnostic marker for lung cancer. Clin Lung Cancer 10, 42-6 (2009).

GLOSSARY

Acylation

Co- or post-translational addition of fatty acyl chain to a protein.

Acetylcholinesterase (AChE)

An enzyme that degrades the neurotransmitter acetylcholine, producing choline and an acetate group, mainly found in central nervous system.

Alzheimer's disease

Neurodegenerative disease characterized by pathological accumulation of amyloid-β-peptide plaques in the brain.

Anomeric

The α (axial) or β (equatorial) stereochemistry description of the glycosidic linkages between two sugars.

Antibodies

Gamma-globulin proteins found in blood or other bodilyfluids of vertebrates, used by the immune system to identify and neutralize foreign objects, such as bacteria and viruses.

Atomic force microscopy (AFM)

A very high-resolution type of scanning probe microscopy used to scan flat media to produce a 3-dimensional profile of the surface. A very sharp tip with a radius of curvature in the nm range (probe), attached to the end of a flexible cantilever, is drawn across the surface. The result is a 3-dimensional contour map of the surface, with resolution in the z-direction of 1-2 Å, and in the x-y plane of 5 Å.

Biologics

Biologically-derived components, products or biochemicals of cells and organisms, or the organisms themselves, which are typically developed to display medicinal properties.

Biomimetics

The description of emulating a natural process, biological structure or function.

Bioorthogonal

A chemically-based process, method or functionality that is compatible and discreet from biologically-based functionality and living processes.

Brewster angle

When light strikes a reflective surface at this angle, the reflected light is plane-polarized.

Brewster angle microscopy (BAM)

A non-invasive technique that can be used to characterize lipid monolayers at the air/water interface, with a typical thickness of ~ 2 nm. The Brewster angle microscope makes use of the fact that an air/water interface does not reflect vertically linearly polarized light that is incident at the Brewster angle, so the water surface appears black. When a lipid monolayer with a different refractive index is spread on the air-water interface a small amount of light is reflected, which can be used to form a high contrast image of the lateral morphology of the monolayer surface.

Caveolin

Lipid raft resident protein involved in membrane structure and endocytosis.

Click chemistry

A collection of chemical methods coined by K. Barry Sharpless able to be applied in biological environments that includes the 1,3-dipolar addition of azides onto alkynes, the Staudinger ligation and thiol conjugate additions to maleimides.

Complement	A biochemical cascade that helps clear pathogens from an organism. A part of the immune system Complement - A part of the innate immune system, the first line of defence.
Deprotection	The process of removing a protecting group to reveal a reactive functionality.
Detergent-resistant membrane domains (DRMs)	Sterol and sphingolipid-rich lipid fractions obtained from detergent extraction of cells.
Diagnostic ion	Product ion whose formation reveals structural or compositional information of its precursor.
Electrospray ionisation (ESI)	A process in which ionized species in the gas phase are produced from a solution via highly charged fine droplets, by means of spraying the solution from a narrow-bore needle tip at atmospheric pressure in the presence of a high electric field.
En bloc	As a whole, all at once.
Endocytosis	Vesicle-mediated uptake of compounds inside the cell.
Envelope	A cell derived, lipid bilayer shell surrounding enveloped viruses.
Enveloped virus	A virus carrying an envelope.
Exocytosis	Release of secretory vesicles from a cell.
Exosome	Lipid bilayer vesicle released from cells by means of exocytosis (40-100 nm).
Fixation	A chemical process by which biological tissues are preserved from decay
Fluorescence resonance energy transfer (FRET)	A technique for investigating a variety of biological phenomena that produce changes in molecular proximity.
Fourier transform infrared reflection/absorption spectroscopy (FT-IRRAS)	A technique allowing collection of infra-red (IR) spectra from molecules in a thin film or monolayer on either a water or metal surface. Infra-red spectra of proteins can identify amide bonds and establish whether the amino acid residues are present in α-helices or β-strands, as well as being able to detect changes in secondary structure.
Ganglioside	A compound of the cell plasma membrane composed of a glycosphingolipid (ceramide and oligosaccharide) with one or more sialic acids linked on the sugar chain. They modulate cell signal transduction and are abundant in lipid rafts.
Giant unilamellar vesicle (GUV)	A large, synthetic lipid bilayer vesicle produced for membrane studies (10-100 μm).
Glycan	A carbohydrate component or molecule composed of multiple sugars. A polysaccharide or oligosaccharide chain with or without branching components.
Glycoconjugates	Glycosylated versions and variants of nucleic acids, proteins, carbohydrates, and lipids.

Monosialotetrahexosylganglio-side (GM1)	A member of the ganglio series of gangliosides which contain one sialic acid residue. Important in neuronal plasticity and repair mechanisms.
GPI predictor	Bioinformatic tool that based on the amino acid sequence predicts whether a protein is likely to be a GPI-anchored protein when mature.
Immunoglobulin E (IgE)	A class of antibody.
Immunohistochemistry	The process of localizing proteins or other molecules in cells of a tissue section exploiting the principle of antibodies binding specifically to antigens in biological tissues.
In-gel digestion	Part of the sample preparation for the mass spectrometric identification of proteins in course of proteomic analysis. It primarily comprises the four steps distaining, reduction and alkylation of the cysteines in the protein, proteolytic cleavage of the protein and extraction of the generated peptides.
In silico	Performed on computer or via computer simulation.
Ion	An atomic, molecular or radical species with an unbalanced electrical charge.
Langmuir trough	Used to compress 2-dimensional monolayers of molecules on the surface of a water subphase, and measure surface phenomena due to this compression. Monolayers can also be transferred from the air-water interface of Langmuir troughs onto solid supports as mono- or multi-layer films (Langmuir-Blodgett films) for further investigation.
LC MS/MS	Liquid chromatography coupled on-line to a tandem mass spectrometer.
Ligation	The process of chemically or biologically connecting two molecules together.
Lipidation	The process of adding a lipid to a molecule.
Lipid raft(s)	Sphingolipid and cholesterol enriched membrane microdomain(s).
Liposome	A synthetic lipid bilayer vesicle produced by mixing phospholipids (40-100 nm).
Mass spectrometer	An instrument that measures the mass-to-charge ratio (m/z) and relative abundances of ions.
Matrix assisted laser desorption ionisation (MALDI)	Formation of gas-phase ions from molecules which are present in a solid matrix that is irradiated with a pulsed laser.
Membrane microdomain	Another definition for lipid raft.
Metabolic engineering	The process of incorporating a chemically modified molecule into biological products using the natural metabolisms and biosynthetic processes of organisms and cells.
Modification specific proteomics	A compendium of biochemical methodologies devoted to the analysis of post translational modifications.

Multivesicular bodies	Vesicles produced inside the cell by spontaneous invaginations of the membrane.
Myoblast	A type of progenitor cell that gives rise to myocytes (skeletal muscle cells) and muscle fibers.
Natural products	Naturally derived secondary metabolites of organisms.
Neural cell adhesion molecule (NCAM)	A binding glycoprotein expressed on the surface of neurons, glia, skeletal muscle and natural killer cells. Has a role in cell-cell adhesion, neurite outgrowth, synaptic plasticity, and learning and memory.
Neural tube	Embryo's precursor to the central nervous system.
Omega-site (ω-site)	The specific amino acid to which the GPI anchor is attached, thus becoming the C termini of the mature protein.
Organellar subproteomics	Proteomics methodologies focused on the analysis of cellular organelles.
Palmitoylation	The covalent attachment of fatty acids, such as palmitic acid, to cysteine residues of membrane proteins.
Parkinson's disease	Neurogenerative disease manifested by the destruction of dopamine-producing neurons in the brain leading to a decrease in dopamine production and signaling.
Paroxysmal nocturnal hemoglobinuria (PNH)	A disease characterised by the absence of cmplement regulatory, GPI-anchored proteins from erythrocytes, caused by a mutation of the enzyme catalyzing the first step in the synthesis of glycosylphosphatidylinositol.
PEGylation	The process by which the polymer PEG is used to increase biocompatibility of molecules.
Prion disease	Pathological conversion of the cellular glycoprotein PrP^c resulting in encephalopathy.
Protecting group	A chemical assembly that is added to a molecule in order to block or deactivate the reactivity of a particular functional group, like alcohols and amines.
Pseudodisaccharide	A sugar conjugate of a sugar-like molecule, like inositol.
Stereochemistry	The defined, geometric or three-dimensional spatial arrangement of chemical groups or atoms of a molecule in space.
Stereoselective	A chemical method that gives preference for a particular stereochemistry in a molecule.
Synthesis	The process of making a molecule in the laboratory using biological and/or chemical methods.
Systemic lupus erythematosus (SLE)	Autoimmune disease characterized by the development of autoantibodies against nuclear antigens and abnormal T-cell signalling, subsequently leading to inflammatory damage of multiple organs.

Tandem mass spectrometer

An instrumental arrangement in which ions are subjected to two or more sequential stages (which may be separated spatially or temporally) of analysis according to the mass to charge (m/z) ratio.

Total synthesis

The complete and chemically precise synthesis of naturally or biologically derived molecules from smaller starting materials.

Viral vector (VV)

A modified virus, designed to deliver therapeutic genetic information to cells.

Virus-like particle (VLP)

A non-infectious virus particle containing no viral genetic material.

INDEX

Acetylation 59

Acylation 4,35,43,58,98

Acetylcholinesterase (AChE) 19,20,23,58,98

Aerolysin 19

Alzheimer's disease 1,8,12,98

Alkaline phosphatase 19,20,22,35,39,57

Angiotensin-converting enzyme 21-23

Anomeric 67,98

Antibody, antibodies 3,6,7,21,26,27,41-46,57,64-67,73-75,87,89-92,98,100

Atomic force microscopy (AFM) 34,38-40,98

Biologics 64,65,74,98

Biomimetic 37,74,98

Biosynthesis, synthesis 3,4,9,10,19-22,43,64-69,70,72,75,101,102

Bioorthogonal 71, 72, 98

Brewster angle 37,98

Brewster angle microscopy (BAM) 37,38,98

Carboxypeptidase M 20,23,57

Carcinoembryonic antigen 20,23,43

Caveolin 1-6,8-11,98

CD4 7,25

CD52 23,46,65-67

CD55 20,21,23-25,83,87

CD59 9,20,21,23-25,57,83-85,91-93

Cholesterol 1-5,8,10,12,19,21,23,34,36,38-41,45,47,55,88,89,100

Cleavage 1,8,21-23,34,35,42-46,90,100

Click chemistry 69,71-73,75

Complement 9,19,21,24,25,84,86,87,89,99

Complement regulator 9,19,21,24,25,84,86,87,89,99

Complement-regulatory proteins 9,19,21,24,25,84,86,87,89,99

Deprotection 67,99

Detergent 1-3,19,21,22,26,27,29,34-37,39,43,46,53,55,56,87,92,99

Diagnostic ion 53,60,61,99

Ectoenzyme 34

Electrospray ionisation (ESI) 56,59,60,99

Embryogenesis 19

Embryonic lethality 19,65,92

Endocytosis 5,6,8,70,98

Endoplasmic reticulum 4,10,20,22,53,56

Envelope 25,72,73,83-87,99

Enveloped virus 83-85,89,99

Erythrocyte 1,19,23-25,87-89,101

Exocytosis 5,90,99

Exosome 23,90,93,99

Fixation 26,99

Fluorescein 71

Fluorescence resonance energy transfer (FRET) 3,21,42,99

Fourier transform infrared reflection/absorption spectroscopy (FT-IRRAS) 39,99

Gangliosides 8,12,19,25-27,45,100

GFP 41,70-71,88,91,92,94

Giant unilamellar vesicle (GUV) 39,89,90,93,99

Glossary 98-102

Glycan 4,21,41,43,44,53,54,58,60,64-74,99

Glycoconjugates 20,64,99

Glycophospholipids 64

Glycosylation 43,59

GM1 (monosialotetrahexosylganglioside) 7,8,21,41,100

GPI biosynthesis 20,67

GPI predictor 61,100

GPI signal sequence (GSS) 4,56,87

GPI-specific phospholipases 19,21,22,24,34,35,37,42-47,53,55,56,58-60

GPI-specific phospholipase C 21,22,24,35,43,46,53,55,56,58

GPI-specific phospholipase D 21,22,37,55,56,58

Green fluorescent protein (GFP) 41,71,88,91,92,94

Haemogglutinin 74

Human immunodeficiency virus type 1 (HIV-1) 72-74,83-85

Immunoglobulin 72,100

Immunoglobulin E (IgE) 21,100

Immunohistochemistry 26,100

Immunostaining 26

In-gel digestion 56,100

In silico 61,100

Langmuir trough 37,100

LC MS/MS 55,56,61,100

Leaflet, inner 1,3,4,35,36,40

Leaflet, outer 1,3,4,19,21,23,24,26,35-37,40,44,45,88

Ligation 69-72,98,100

Lipase 22

Lipidation 69,100
Lipid raft(s) 1-11,19,21,24-26,34,35,38,
39,41,47,54,55,68,83,88,89,98,99,100
Liposome 24,25,35,36,37,44,45,47,89,90,92,93,100

Magnetic 46,75,86,90,92,93
Matrix assisted laser desorption ionisation
(MALDI) 27,59,60,100
Membrane microdomain 1,19,21,45,83,84,88,100
Membrane vesicle 5,23,24,45,90,92,93
Metabolic engineering 71,73,100
Modification specific 53-55,60,100
Microparticle 90,91,93
Model membrane 2,3,8,12,34
Monosaccharide 71
Monosialotetrahexosylganglioside (GM1)
7,8,21,41,100
Multivesicular bodies 23,90,101
Myoblast 9,23,101

Nanoparticle 75,89-93
Natural products 64,101
Neural Cell Adhesion Molecule (NCAM)
8,20,23,101
Neural tube 19,101

Omega-site (ω-site) 53,54,56-58,61,101
Organellar subproteomics 54,101

Palmitoylation 88,101
Parkinson's disease 8,12,101
Paroxysmal nocturnal hemoglobinuria (PNH)
9,21,24,89,101
PEGylation 89,101
Phosphatidylinositol 4,20,21,24,34,43,65

Phosphorylation 7,8
Phospholipases 19,21,22,24,34,35,37,42-
47,53,55,56,58-60
Prion disease 1,8,24,101
Prion protein 21,23,24,66,69
Prostasome 23,25
Protease 8,22,23,43,46,55,69
Protecting group 67,99,101
Pseudodisaccharide 66,67,101

Receptor 4,7,12,19,20,22,23,28,40,41,44,46,53,
55,72,87,88,91,93
Release 19,22-24,26,34,46,58,69,72,99
Retrovirus 25,83

Seminal plasma 23,24,
Shedding and uptake 19,23,25-27,28
Signal transduction 1,4,5,8,12,19,21,22,26,39,43,65,
67,68,88,99
Stereochemistry 98,101
Stereoselective 87,101
Synthesis, biosynthesis 3,4,9,10,19-22,43,64-
69,70,72,75,101,102
Systemic Lupus erythematosus 7,101

Tandem mass spectrometer 56,60,100,102
Total synthesis 64,67,68,102
Transmembrane proteins 20,34,40,41
Triton X-100 2,21,26,27
Uptake 5,6,19,23-28,69,99
Urokinase-type plasminogen activator 23,44

Variant surface glycoprotein (VSG) 19,24,41,43,65
Viral vector (VV) 83-87,89,91,92,102
Virus-like particle (VLP) 83-87,90,102

9 781608 053759